MICROTECHNOLOGY AND MEMS

MICROTECHNOLOGY AND MEMS

Series Editor: H. Fujita D. Liepmann

The series Microtechnology and MEMS comprises text books, monographs, and state-of-the-art reports in the very active field of microsystems and microtechnology. Written by leading physicists and engineers, the books describe the basic science, device design, and applications. They will appeal to researchers, engineers, and advanced students.

Mechanical Microsensors
By M. Elwenspoek and R. Wiegerink

CMOS Cantilever Sensor Systems
Atomic Force Microscopy and Gas Sensing Applications
By D. Lange, O. Brand, and H. Baltes

Micromachines as Tools for Nanotechnology
Editor: H. Fujita

Modelling of Microfabrication Systems
By R. Nassar and W. Dai

Laser Diode Microsystems
By H. Zappe

Silicon Microchannel Heat Sinks
Theories and Phenomena
By L. Zhang, K.E. Goodson, and T.W. Kenny

Shape Memory Microactuators
By M. Kohl

Force Sensors for Microelectronic Packaging Applications
By J. Schwizer, M. Mayer and O. Brand

Integrated Chemical Microsensor Systems in CMOS Technology
By A. Hierlemann

CCD Image Sensors in Deep-Ultraviolet
Degradation Behavior and Damage Mechanisms
By F.M. Li and A. Nathan

Micromechanical Photonics
By H. Ukita

Fast Simulation of Electro-Thermal MEMS
Efficient Dynamic Compact Models
By T. Bechtold, E.B. Rudnyi, and J.G. Korvink

T. Bechtold E.B. Rudnyi J.G. Korvink

Fast Simulation of Electro-Thermal MEMS

Efficient Dynamic Compact Models

With 108 Figures, 8 in Color and 9 Tables

Springer

Dr. Tamara Bechtold
Dr. Evgenii B. Rudnyi
Professor Dr. Jan G. Korvink
Universität Freiburg
Institut für Mikrosystemtechnik
Georges-Köhler-Allee 103
79110 Freiburg, Germany
E-mail: Tamara.Bechtold@imtek.uni-freiburg.de
rudnyi@imtek.de
korvink@imtek.de

Series Editors:

Professor Dr. Hiroyuki Fujita
University of Tokyo, Institute of Industrial Science
4-6-1 Komaba, Meguro-ku, Tokyo 153-8505, Japan

Professor Dr. Dorian Liepmann
University of California, Department of Bioengineering
6117 Echteverry Hall, Berkeley, CA 94720-1740, USA

ISSN 1615-8326

ISBN-10 3-540-34612-0 Springer Berlin Heidelberg New York

ISBN-13 978-3-540-34612-8 Springer Berlin Heidelberg New York

Library of Congress Control Number: 2006930117

Springer is a part of Springer Science+Business Media.

springer.com

Typesetting by the authors and SPi using a Springer LaTeX macro package
Cover concept: eStudio Calamar Steinen
Cover production: *design & production* GmbH, Heidelberg

Printed on acid-free paper SPIN 11600572 57/3100/SPi - 5 4 3 2 1 0

Dedication

TB: For Dennis

Preface

In 2001, an electrical engineer, a chemist and a mechanical engineer needed to solve a common problem: to develop a compact electro-thermal model for system-level simulation. An initial idea was to use the conventional route of compact thermal modeling but fortunately we found a much better alternative: formal model order reduction. This book presents results of our research and an introduction to both electro-thermal simulation and model order reduction.

We are practitioners and the book is written for practitioners as well. A background in mathematics is required but the main emphasis of the book is on the original ideas. There are many practical examples that show how one can use model reduction to build an accurate dynamic compact model in an automatic fashion directly from a finite element model.

When we started this work, there was almost no software to use modern model reduction in practical work. Over time we have developed a software environment that allows us to start from a thermal model developed in ANSYS®, perform model reduction directly for an ANSYS model, and then simulate the reduced model and visualize the results. The software is described in the book and available from our Web site. Even if the background mathematics happens to be too difficult to comprehend, a reader will still be able to perform model reduction of their own models.

Model reduction is quite a general technique and it is not limited to thermal problems. On the other hand, it happens that a thermal problem is the most natural problem for the application of model reduction. As such, the book will be a good introduction to model reduction for any engineer even though its main focus is electro-thermal simulation.

Acknowledgments

The book was written while working in the Simulation Lab of Professor Korvink, and we would like to thank all our colleagues for the nice working atmosphere.

The original engineering problem occured on the European project Micropyros. We appreciate the help of our colleagues to understand the problem, and acknowledge the financial support of EU.

We thank Professor Lohmann and his team for many meetings both in Freiburg in Bremen where we discussed different aspects of model reduction and its application in engineering. We acknowledge the financial support of DFG for our joint research.

Finally, we thank IOP for permission to use material in this book from three of our papers published in *Journal of Micromechanics and Microengineering* as follows:

T. Bechtold, E.B. Rudnyi, J.G. Korvink. Error indicators for fully automatic extraction of heat-transfer macromodels for MEMS. *Journal of Micromechanics and Microengineering*, 2005, vol. 15, N 3, pp. 430–440.

T. Bechtold, E.B. Rudnyi, Markus Graf, Andreas Hierlemann, J.G. Korvink. Connecting heat transfer macromodels for MEMS array structures. *Journal of Micromechanics and Microengineering*, 2005 v. 15, N 6, pp. 1205–1214.

T. Bechtold, E.B. Rudnyi, J.G. Korvink. Dynamic electro-thermal simulation of microsystems – a review. *Journal of Micromechanics and Microengineering*, 2005, v. 15, N 11, R17–R31.

Freiburg, 2006 *The authors*

Contents

1

Introduction

This text addresses the application of a modern mathematical approach, referred to as model order reduction (MOR), for efficient simulation of electro-thermal microsystems. It is written for engineers that use high-dimensional finite element models (alternatively, spatial discretization can be used as well) during device simulation. It describes the automatic generation of accurate dynamic compact thermal models directly from finite element models. Such compact models are convertible into hardware description language form and can be directly used in system-level simulation or employed for extremely effective design optimization of electro-thermal microsystems.

In general, the design of microsystem devices often depends on large-scale transient simulation of coupled physical domains, such as thermal, mechanical, electrical, etc. This requires the solution of very large systems of ordinary differential equations (ODEs), resulting from the spatial discretization of a computational domain. However, instead of a common "brute force" approach to integrate a large system of ODEs, one can use modern mathematical methods to drastically reduce the problem dimension, and thereby achieve dramatic speedup of the calculation time. Hence, nowadays it is possible to simulate models that only several years ago were too large (due to lack of time, computer memory or computer speed). Indeed, it has been shown that for many MEMS devices, such as accelerometers, gyroscopes and many different electro-thermal devices, the number of ODEs obtained from finite element modeling can be reduced by several orders of magnitude almost without sacrificing precision.

This book describes a complete MOR methodology and software environment at the engineering level. It is equipped with a large number of practical examples, to show readers how to considerably speed up simulation in a concrete problem. Although the model order reduction approach can be applied to different physical domains, in this book we focus on electro-thermal MEMS.

1.1 MEMS, Compact Modeling and Model Order Reduction

The development of increasingly complex microstructures (in the following we will call all microsystems MEMS[1], even if functionality other than micro-electromechanical is employed) demands sophisticated simulation techniques for design, control and optimization [1, 2]. Often, system-level simulation, which includes several single devices placed on a chip together with their driving circuitry, is indispensable. Although no universal simulation strategy currently exists to cover all MEMS design situations [3], reduction of the problem size drastically reduces the computational work. This book is an overview on how to automatically produce reduced models for system-level simulation of MEMS, using modern mathematical approaches.

Traditionally size reduction is performed via compact modeling, which was developed in electrical engineering long before MEMS. The goal of compact modeling is to create a small size equivalent network of resistors, capacitors, inductors, etc. which accurately describes the dynamics of the device and can be directly inserted into SPICE-like simulators. Naturally, MEMS engineers try to use the same methodology. Mathematically speaking, compact modeling starts by choosing the topology of a small-dimensional equivalent circuit (see Figure 1.1). During the second step, parameters within this network (resistivities and capacitances) are found by fitting model parameters to measured or simulated curves. This approach requires the designer to choose the correct network topology intuitively, i.e. without strict guidelines, and then to perform a model parametrization. It should be noted that although the second step requires time-consuming data fitting, the first step usually takes even more time in practice as it is based on intuition.

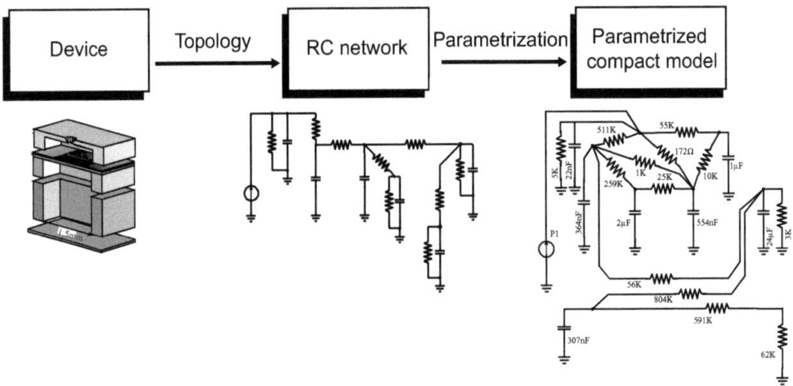

Fig. 1.1. Compact modeling flow of the example thermal MEMS model. RC network pictures courtesy of M. Salleras (UB, Spain).

[1]MEMS traditionally stands for micro-electromechanical systems.

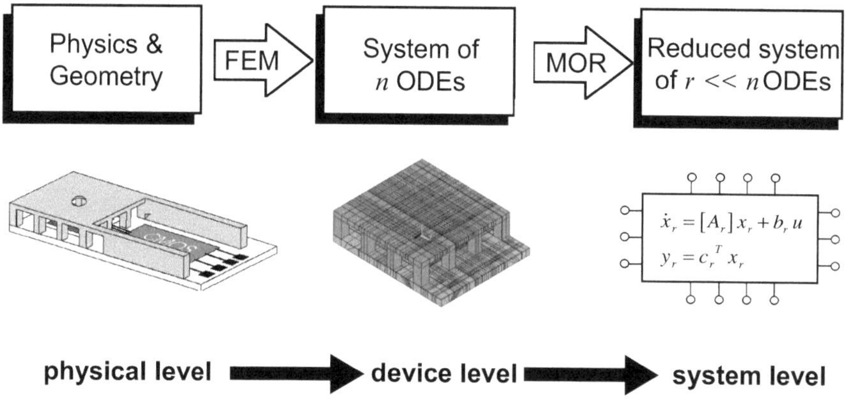

Fig. 1.2. Model order reduction: a switch from device to system level simulation.

An alternative to compact modeling is mathematical model order reduction (also called the approximation of large-scale dynamic systems [4]). Figure 1.2 shows the MOR flow. The simulation of a single device starts with the governing partial differential equations (PDEs). One example is a heat transfer PDE, Eq. 2.5, which takes a central place in this book. The next step is the discretization in space of the original PDE, using, for example, the finite element method (FEM), which integrates the PDE over a number of small nonoverlapping subsets of the complete domain. This results in a system of ordinary differential equations, whose dimension is proportional to the number of introduced nodes. The finer the spatial discretization required, the more nodes are produced. Due to the complex nature of MEMS, its discrete models are usually large (100,000 equations are the engineering standard nowadays). The second step is mathematical model order reduction, which is based on the transformation of a high-dimensional system of ODEs to a low-dimensional one, done by projection. This conversion is formal, robust and can be fully automated. If necessary, the reduced system can also be represented as an equivalent electrical circuit and inserted into a system-level simulator. Hence, mathematical model order reduction can be considered as "compact modeling on demand" [5].

In Table 1.1 the most important properties of compact modeling and mathematical model order reduction are contrasted. Compact modeling requires the designer to intuitively choose the topology of a small-sized equivalent network, which is not a trivial task. It further requires either simulation of the original large-scale system or the use of experimental data. These are then used for the parametrization (via data fitting) of the compact model. No mathematical properties of the original system matrices, or their connection to the matrices of the reduced ODE system are taken into account. Model order reduction, on the other hand, does not require simulation of the original system. Instead, it reduces the original large-scale system matrices using the concept of mathematical projection (explained in detail in Chapter 3). Hence, the reduced system is obtained completely formally and without relying on intuition.

Table 1.1. Compact modeling vs. model order reduction (MOR)

Method properties	Compact modeling	MOR
reduced model topology	obtained by intuition	formally obtained
simulation of the original model	necessary	not necessary
parameter extraction	necessary	not necessary
experimental results	can be used	cannot be used
system matrices of the original model	not used	used

In this book, we focus on MOR, because for linear systems and systems with nonlinear input functions[2] it is much more effective than compact modeling. Presently, the development of model order reduction is strongly driven by diverse MEMS engineering applications [6, 7] that would benefit significantly if the dimension reduction could be done in a completely automatic fashion. In general, one can say that the development of efficient MOR methods for automatically creating accurate low-order dynamic models is about to become a major subject of MEMS simulation and modeling research.

1.2 MEMS and Electro-thermal Simulation

Modeling of electro-thermal processes, e.g. Joule heating, becomes increasingly important during microsystems development [8]. For example, with the decreasing size and growing complexity of micro electronic systems, the power dissipation of integrated circuits has become a critical concern. The thermal influence upon the device caused by each transistor self-heating and the thermal interaction with tightly placed neighboring devices cannot be neglected because excessive temperatures change transistor characteristics and may cause malfunction or even destruction of the device. Whereas Joule heating in microelectronics is a "parasitic" effect, many MEMS devices, such as microsensors and microactuators, use it (directly or indirectly) as a functioning principle. For example, microhotplate-based devices can use Joule heating to achieve the operating temperature. Micromechanical devices with electro-thermal actuation can use Joule heating for electro-thermal expansion. Microfluidic devices employ Joule heating to expel microdroplets out of microfabricated reservoirs, etc. In all design situations above, the engineer's task is to predict the temperature distribution for the given electrical input and the impact of temperature on the device electronics in turn, i.e. to run an electro-thermal simulation. To go a step further, in each sequence of joint electro-thermal simulations the temperature field is computed on a discrete grid whose size, as already mentioned, easily exceeds 100 000 degrees of freedom, i.e. ordinary differential equations.

When we started this research in 2001, MOR approaches were only used for electric circuit interconnects and structural mechanical problems. No efforts to apply MOR to electro-thermal MEMS models have been made so far. At the same

[2]Here we mean the dependence of the input function on state variables, such as the dependence of heater resistivity on temperature.

time, thermal problems, which after spatial discretization result in first-order ODE systems (see Eq. 2.14), are the "most convenient" case from the MOR point of view. Consequently, they provide the best examples to learn MOR and will be targeted in this book.

1.3 Thematic Outline

The purpose of this text is to guide MEMS engineers to apply MOR for efficient simulation of electro-thermal models or electro-thermal domains of coupled models. Along the way, the reader will develop a deeper understanding of

- electro-thermal modeling in general,
- how mathematical MOR works,
- how it can be applied to electro-thermal systems and to coupled systems which include thermal domains,
- how reduced order models can be used within system-level simulation or within a design optimization process.

We thus begin in Chapter 2 with an overview of dynamic electro-thermal simulation of microsystems. The solution of the heat transfer partial differential equation, Eq. 2.5, which is central to electro-thermal microsystem simulation, is discussed. Different levels of solution approaches are presented, starting from the most rigorous, when the heat transfer can only be modeled by a set of coupled partial differential equations, and finishing with the final level of approximation, which is compact thermal modeling.

In Chapter 3 we present the most important algorithms for MOR of linear first-order ODE systems. All MOR methods are based on the projection of a large-scale system to a low-dimensional subspace. Different projections characterize different methods. The central focus of this tutorial is the Arnoldi algorithm, which is based on the projection to a so-called Krylov subspace. We will also discuss the mathematically optimal control theory methods [9, 10], which already have a long tradition in MOR of linear systems of moderate size, and the commercially available Guyan algorithm [11], which up to now has been mostly used for mechanical engineering problems.

In Chapter 4 we describe the available software environment and give a practical guideline how to use it. The software tool *MOR for ANSYS* (pronounced "more for ANSYS) has been developed during the last five years at the laboratory in the simulation. Department of Microsystem Engineering, University of Freiburg (see [12]). It implements the Arnoldi algorithm to generate compact models directly from ANSYS models. Apart from electro-thermal models *MOR for ANSYS* can also be used for the reduction of second-order systems, i.e. for structural models. We also guide the reader in how to perform the substructuring in ANSYS, which is based on the Guyan algorithm (described in Chapter 3) and can be used for the MOR of electro-thermal models as well, although it originally targeted non-damped mechanical models. We further describe how to use the SLICOT library (see [13]), which implements the most important control theory methods and can be applied for MOR of arbitrary first-order ODE systems of moderate size.

In Chapter 5 we describe three novel MEMS devices used as case studies for MOR. These are a pyrotechnical microthruster, a thermally tunable optical filter and a microhotplate gas sensor. The formal switch from electro-thermal to "pure" thermal modeling, under the assumption of homogenous heat generation within the lumped resistor, is discussed as well. We further present the numerical results of model reduction for the three case studies and consider different important aspects of the Arnoldi algorithm, such as the approximation of the complete output and the reduction of systems with nonlinear input functions. Results of Arnoldi-based reduction are compared with the results of control theory methods and those of substructuring. The setup for the system-level simulation (using SABER) and the solution of the inverse thermal problem via MOR are presented in this chapter as well.

While Chapters 3 to 5 address the basic level of MOR that can be used in practice, Chapter 6 outlines advanced topics. These are the error estimate for MOR, MOR of interconnected MEMS systems, parametric MOR, MOR of nonlinear and second-order systems and methods based on low-rank Gramian approximation.

2

Dynamic Electro-thermal Simulation of Microsystems

Thermal management has become a crucial issue in designing modern microelectronic and microelectromechanical components and systems. Only after a careful study of thermal effects in microsystems can we predict their performance, reliability and yield.

In integrated circuits, thermal management is important on all levels, starting from the transistor level, over the chip level, the package level and the printed circuit board (PCB) level, to the system level. The self-heating of transistors may lead to oscillation phenomena [14] affecting the whole circuit behavior. High current during electrostatic discharge may increase the local temperature too much and destroy the transistor. Self-heating of an interconnect may reduce its average time to failure. It is important to position elements on a chip in such a way that the effect of heating is minimal with respect to the chip's functioning [15]. Efficient heat transfer is of paramount importance for packaging [16, 17] as the reliability depends heavily on thermal management [18–21]. Finally, the placement of semiconductor devices on a PCB should also be optimal from the thermal management viewpoint. It is worthy of note that the reliability of the whole board is tied to its temperature regime because there are many mechanisms that lead to enhanced board degradation at elevated temperatures.

The same issues concern microsystems [22] although the requirements here may be different. If for conventional integrated circuits the goal is to keep temperature close to room temperature, in microsystems one needs to keep some parts at relatively high temperatures; usually several hundred degrees. Examples can be found in numerous MEMS devices, such as thermal actuators [23], thermal flow sensors [24], microhotplate gas sensors [25, 26], tunable optical filters [27–29] and many others. The heating changes optical properties in optical filters, reduces activation energies to allow metal oxides to detect gases and causes mechanical stress and thus movement in actuators. The heat transfer through moving fluid allows us to measure the flow rate in heat-flow sensors.

The diverse engineering problems mentioned above have in common that an engineer should be able to predict the temperature distribution for the given electrical input and estimate how temperature distribution in turn affects the electrical part. In other words, he must run a joint electro-thermal simulation [30]. As for simulation of an electrical domain, there are readily available tools, so the real task becomes one of enhancing them in order to take into account a variety of electro-thermal and thermo-electric effects. Hence, solution of the heat transfer equation plays a central role in electro-thermal simulation.

The main goal of this chapter is to present a unified description of different levels of solution approaches for the heat transfer partial differential equation.

Of course, in many cases the coupling of the heat transfer equation to other physical domains must be taken into account as well. Throughout this chapter, we will briefly review different couplings and discuss what is necessary to decouple the heat transfer equation in order to be able to solve it separately.

Section 2.1 gives a short overview of different electro-thermal and thermo-electric effects in microsystems. Throughout this book we will take into account only the Joule heating effect, although the MOR approach can be used for other effects as well. The importance of the Joule heating effect for MEMS and microelectronic systems will be highlighted through examples in Section 2.2. Sections 2.3 and 2.4 describe physical modeling with the heat transfer equation and its couplings to other physical domains at the device level. Throughout Section 2.5 we discuss how to solve the parabolic heat transfer equation on its own. The first topic is the linearization of the heat transfer equation, because the solution of the linear equation is much faster than that of the nonlinear one. Analytical solutions are briefly reviewed. We consider further the most frequent approach, which is based on spatial discretization of the thermal domain (often called the brute force method). The final level of approximation, dynamic compact thermal modeling is described in Section 2.6. It allows us to perform effectively transient electro-thermal simulations at the system level. We include here the conventional, nonautomatic approaches, such as the RC-ladder network, semi-automatic approaches such as modal approximation, as well as the increasingly popular MOR approach, which can be made fully automatic. Finally, Section 2.7 concludes the chapter by displaying a schematics of the modeling approaches discussed (see Figure 2.12).

2.1 Overview of Coupled Electro-thermal and Thermo-electric Effects

A classification of the most important electro-thermal effects in semiconductor and IC-related materials utilized in microsystems is given in Table 2.1. A brief explanation of each effect is presented below (SI units are used).

Heat conduction at the macroscopic level means that when a temperature gradient exists within a solid body, heat energy will flow from the region of higher temperature to the region of lower temperature. This phenomenon is described by the Fourier law:

$$q = -\kappa \nabla T \tag{2.1}$$

Table 2.1. Electro-thermal signal conversion effects in microsystems (see also Natan 1999)

	Thermal	Electrical
Thermal	heat conduction	thermoresistance Seebeck effect pyroelectricity
Electrical	*Joule heating* Peltier effect Thomson effect	electrical conduction

This equation determines the heat flux vector (the amount of heat penetrating a unit of area per unit of time) in W/m^2 for a given temperature profile T and thermal conductivity κ of the body (a material property that describes the rate at which heat flows within a body for a given temperature difference) in W/(mK). All quantities relate to a specific point. The minus sign ensures that heat flows down the temperature gradient.

Electrical conduction describes the electrical current flow in the presence of the electrical potential gradient. If, in Equation 2.1, the temperature is replaced by the electric potential φ and the heat flux vector by the electric current density vector j, the solution of a corresponding problem of electric conduction is given by

$$j = -\sigma \nabla \varphi \tag{2.2}$$

where σ is the specific electric conductivity in S/m.

Joule heating is the dominant mechanism for heat generation due to the flow of electrical current through the material. It is defined by Joule's law, which for an ohmic conductor has the form:

$$Q = j^2 \rho \tag{2.3}$$

where j is the current density vector in A/m^2, ρ is the specific electric resistivity in Ωm and Q is the generated heat per unit volume in W/m^3. This is usually the main effect responsible for heat generation in resistively heated microsystems. A number of applications will be presented in Section 2.2.

The thermoresistance effect states that the specific electric resistivity of the material can be expressed as a function of temperature:

$$\rho(T) = \rho_0(1 + \alpha T + \beta T^2 + \ldots) \tag{2.4}$$

where α and β are the temperature coefficients of the material in K^{-1} and K^{-2} respectively, and ρ_0 is the specific electric resistivity at zero temperature. Positive temperature coefficients are peculiar to pure metals and some alloys. In a number of microsystem applications, metal resistors made of platinum, nickel, copper etc., are used as heating and sensing elements [25, 27]. In such cases Equation 2.4 is on one side of paramount importance for temperature control, and on the other side it contributes to the nonlinearity of the governing heat transfer equation via Equation 2.3.

The Seebeck and Peltier effects are associated with a junction of two different conductors or semiconductors, a so-called thermocouple. If there is a temperature difference maintained between the junction and the free ends, an open-circuit

Seebeck voltage is obtained at the nonconnected end. The Seebeck effect is investigated for use in miniaturized voltage sources [31]. Those devices utilize heat generated by the human body and supply electronic devices (e.g. watches) with a minimum amount of electrical power. It is further used in thermoelectric infrared gas sensors [32]. The Peltier effect describes the generation or absorption (depending on the direction of the current) of heat in a thermocouple when the current flows through the junction in the absence of any temperature gradient. Contrary to Joule heating, the Peltier effect is reversible. It can be effectively used for heat transport opposite the temperature gradient and hence for cooling down the environment. It is mainly used for controlling the temperature of chips while measurements are being made. On-chip integrated Peltier elements are ideally suited for highly localized on-chip thermal stabilization [33].

The Thomson effect is complementary to the Peltier effect. In the Thomson effect two dissimilar materials are not needed but a current passed along a conductor, when a temperature gradient is maintained, results in either absorbance or generation of additional heat (in addition to Joule heat).

The pyroelectric effect describes the change of the electric polarization induced by a change of temperature in some nonlinear dielectric materials, such as PZT-ceramics, PbTiO3, PVC etc. Pyroelectric materials are used within the thermal sensors and infrared detectors. Pyroelectric motion (infrared) sensors are used in a wide variety of applications. They are commonly found in security products, such as burglar alarms, motion detectors and intrusion detection systems [34]. These sensors are also useful in environmental systems, lighting controls, visitor announcers, robotics, and artificial intelligence. They work by detecting the infrared heat emitted by the human body.

2.2 Joule Heating in Microsystems

Joule heating happens to be the dominant mechanism for heat generation in microsystems. Some devices like microsensors and microactuators are designed to optimize the Joule heating effect in a controlled manner in order to improve transduction efficiency. For others, like integrated circuits (ICs), this effect is "parasitic", and hence their design aims at the suppression of Joule heating. Below we review several MEMS devices whose working principle is directly or indirectly based on Joule heating and mention the most common "parasitic" effects due to Joule heating in microelectronic devices and their packaging.

Microhotplate-based devices, such as gas sensors [36–41], optical filters [27], thermal flow sensors [24], solid fuel microthrusters [42], thermal infra-red emitters [43] etc. use Joule heating to realize a desired functionality. Maximum temperatures for the operation of these devices reach several hundred degrees Celsius. In order to significantly reduce the required electrical heating power, the resistive heater structures are placed on a thermally isolated micromachined platform called a membrane (see Figure 2.1). Elevated temperatures may be essential for the onset of operation of the device (e.g. by gas sensors, microthrusters or infra-red emitters) or controlled temperature changes are used for tuning the structure‘s characteristics (e.g. in optical filters).

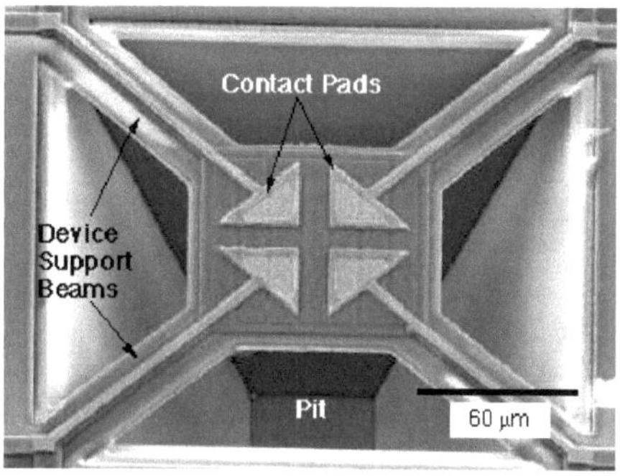

Fig. 2.1. Photomicrograph of a single microhotplate element from [44], with the permission of IEEE.

Micromechanical devices with electro-thermal actuation are used to perform direct mechanical actuation through thermal expansion resulting from Joule heating of selected microstructures. Both in-plane actuators, based on thermally induced extension [46], and out-of-plane actuators [47], based on different thermal expansion coefficients that rely on the thermal expansion of silicon, polysilicon, nickel, or related materials have been extensively studied and used in MEMS structures. Thermal microactuators based on polymers such as polyimide, thermal bimorphs, pseudo-bimorphs, buckling beams, compliant structures and high-aspect-ratio structures have also been investigated (see [44] and the references therein). The working principle of an embedded electro-thermal-compliant actuator is shown in Figure 2.2. The wide arms, which have lower electrical resistance than the narrow arms, draw more current and get hotter. As a result, the structure bends.

Microfluidic devices employ Joule heating to expel microdroplets out of microfabricated reservoirs. In a reservoir, air bubbles are generated within the liquid phase at the bottom of the reservoir that is above the heater, as schematically shown in Figure 2.3. As the properties of microscale bubbles are dominated by surface tension effects, they tend to be very stable and hence useful for generating mechanical

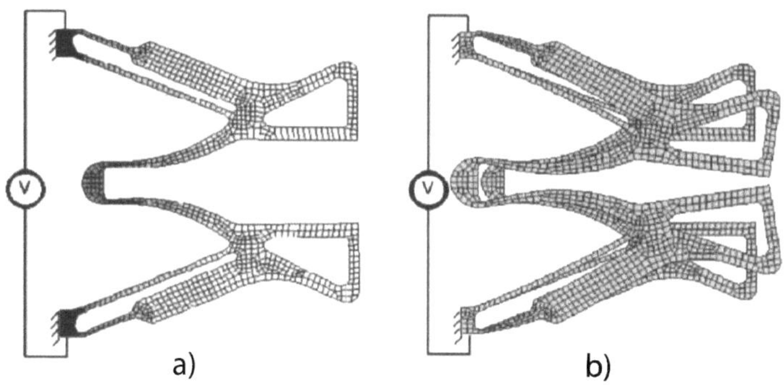

Fig. 2.2. An example of electro-thermal embedded actuation: (a) original configuration; (b) deformed configuration superimposed on the original from [45], with the permission of Elsevier.

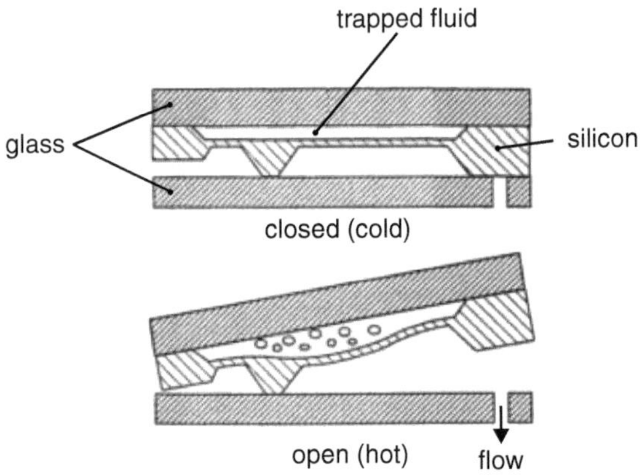

Fig. 2.3. Functioning principle of a thermopneumatic microvalve, adapted from [55].

work. The dramatic increase in volume, which is due to bubble generation, forces small amounts of liquid to leave the chamber through a micronozzle. This technology is used in a broad spectrum of MEMS devices starting from ink jet print heads [48]. The thermopneumatic effect is further used to vaporize a working fluid through Joule heating. The increased gas pressure actuates a diaphragm. This deflection can be used in micropumps and microvalves [49–53] as components of microfluidic systems.

Microelectronic devices and their packaging are significantly influenced by temperature. In contrast to the previously described devices, Joule heating is not a desirable effect in microelectronic systems, but rather a “parasitic” one. Joule heating activates damaging mechanisms such as corrosion of metallization and bondpads, fatigue of wires or bonds, contact spiking, electrostatic discharge, electrical overstress (causing alone over 40% of all failures), electromigration, die thermal breakdown, damage of wire bonded interconnections, flip-chip joints damage, cracking in plastic packaging etc. In addition to the above, there is almost no room for heat dissipation due to the huge number of transistors per unit area (modern VLSI contain 10^4 to 10^7 transistors per chip) and also no longer enough time due to higher switching frequencies. Today’s heat flux densities on the chip easily exceed 100 W/cm^2 [56]. Not only the operating temperature by itself, but also fluctuations are dangerous for the device’s reliability. It was already known in 1989 that a temperature oscillation of only 15 K increases the failure rate by eight times [56]. Due to the fact that temperature gradients are steeper in transients than in the steady-state phase [58], transient (dynamic) thermal stresses are more dangerous. Violent operating conditions include the following dynamic effects: avalanche and electro-static discharge. Further effects such as temperature oscillations, current and voltage thermal drift or temperature gradient over the device [56], may degrade the device performance.

It should be noted that heat transfer modeling on its own varies in importance for the above applications. In microhotplate-based devices it plays a major role and can

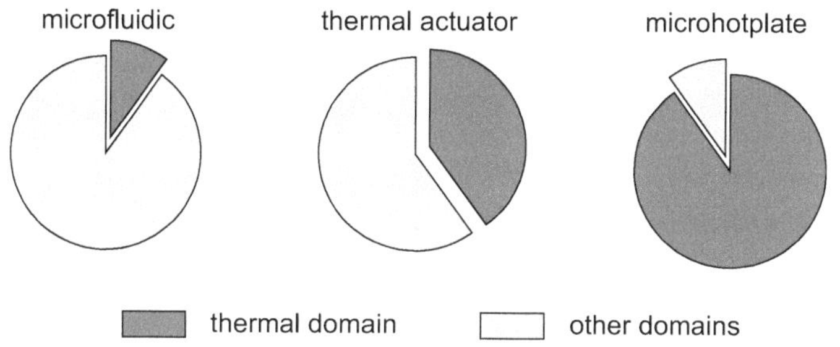

Fig. 2.4. Qualitative importance of heat transfer modeling for different types of MEMS devices.

easily be decoupled from other physical domains. In microfluidic devices, on the other hand, it plays a relatively small role, because it is strongly coupled to Navier–Stokes equations. In Figure 2.4 the qualitative importance of heat transfer modeling for the different types of MEMS devices is schematically presented. However, it is difficult to set clear borders and each engineering problem has to be considered separately.

2.3 Physical Model – Heat Transfer Equation

The most general way to model heat transfer at the device level in crystalline semiconductors is by use of the phonon Boltzmann transport equation: it describes the heat exchange between heat carriers (phonons), which are the energy quanta of lattice vibrations [57] and the lattice. It should be used when the heat carrier mean free path is comparable with the characteristic dimensions at microscale. To give an example, the phonon mean free path in silicon at 300 K is about 300 nm. However, if one can assume that the Boltzmann–Maxwell distribution is valid for any small volume, or in other words that the temperature is defined at any point within the domain, a hyperbolic heat equation is used. It predicts a finite wave speed of heat propagation and is valid at the very small time scale of femtoseconds, for example during laser heating of thin metal films. If we further, assume that the speed of the thermal waves is infinite, the parabolic heat transfer equation can be used instead. Its solution is our main topic, since the assumptions mentioned above are valid for the devices in question.

The parabolic heat transfer equation specifies the complete spatial and time profile of a temperature distribution within a computational domain Ω, limited by the boundary $\partial\Omega$. In solid it has the form:

$$\nabla \bullet (\kappa \nabla T) + Q - \rho C_p \frac{\partial T}{\partial t} = 0 \tag{2.5}$$

where $\kappa(r)$ is the thermal conductivity in W/mK at the position r, $C_p(r)$ is the specific heat capacity (a material property that indicates the amount of energy a body stores for each degree increase in temperature, on a per unit mass basis) in J/kgK,

$\rho(r)$ is the mass density in kg/m^3, $T(r,t)$ is the temperature distribution and $Q(r, t)$ is the heat generation rate per unit volume in W/m^3. Equation 2.5 states that the temperature profile within a body depends upon the rate of its internally-generated heat, its capacity to store some of this heat by raising its temperature and its rate of thermal conduction to its boundaries (where the heat is transferred to the surrounding environment). The solution of Equation 2.5 requires the determination of initial and boundary conditions (BC) for the computational domain Ω. The temperature distribution T_0 at $t = 0$ serves as an initial condition:

$$\forall r \in \Omega,\ t = 0,\ T(r,t) = T_0(r) \tag{2.6}$$

Prescribed temperature distributions on the boundary in time can be modeled by *Dirichlet boundary conditions* as follows:

$$\forall r \in \partial\Omega,\ T(r) = T_{prescribed}(t) \tag{2.7}$$

Quite often, the boundary temperature is assumed to be constant and equal to the temperature of the surroundings and can be set to zero without loss of generality.

Prescribed normal heat flux through the body boundaries $q_\perp$ in time can be modeled by *Neumann boundary conditions* as follows:

$$\forall r \in \partial\Omega,\ q_\perp(r) = q_{prescribed}(t) \tag{2.8}$$

Convective heat transfer takes place when whole subvolumes move from one place at a certain temperature to another at a different temperature. Hence, the convection is created by fluid flow. In many cases, however, it is possible to eliminate the fluid flow from the computational domain by replacing it with a so-called *convection boundary condition*. It assumes that the normal heat flux through the boundary is proportional to the temperature difference between the boundary and the ambient temperature of the adjacent fluid:

$$\forall r \in \partial\Omega,\ q_\perp(r) = h(T - T_{ambient}) \tag{2.9}$$

where h is the heat transfer coefficient, which characterizes the thermal flow as well as the thermal contact between the conducting solid and adjacent fluid with temperature $T_{ambient}$. Again, quite often $T_{ambient} = 0$. It should be noted that the convection boundary condition turns into Neumann for $h = 0$, or into Dirichlet for $h \to \infty$.

The heat radiation mechanism can be modeled by the so called *radiation boundary condition* as follows:

$$\forall r \in \partial\Omega,\qquad q_{emitted}(\boldsymbol{r}) = \varepsilon \cdot \sigma \cdot A(T^4_{surface} - T^4_{ambient}) \tag{2.10}$$

where $q_{emitted}$ is the emitted heat transfer rate (power), ε is the surface emissivity, σ is the Stefan–Boltzmann constant ($5.669 \times 10^{-8}\,\mathrm{m}^2\mathrm{K}^4$) and A is the radiation surface.

It is worthy of note that in practice it is important to be able to solve Equation 2.5 with different boundary conditions, that is, to create a boundary conditions independent (BCI) model [72]. This is possible at the level of detailed solution of Equation 2.5 (Section 2.5.3) but is difficult at the level of compact models (see Section 2.6.1).

2.4 Decoupling the Heat Transfer Equation from Other Physical Domains

In the general case the heat transfer equation, Equation 2.5, is coupled with other partial differential equations already at the device level. Let us consider the most important coupling and discuss what approximations are necessary to decouple the equations in order to be able to efficiently solve Equation 2.5 on its own, preferably via spatial discretization and MOR.

The first coupling comes through the environment, since the device is usually surrounded by moving fluid (either gas or liquid), which serves to remove heat by natural or forced convection. The heat transfer in the moving fluid is generally a part of the solution of the Navier–Stokes equations. They can be coupled with Equation 2.5 in order to describe the overall heat transfer, which makes the resulting set of equations quite complicated. In some cases it is possible to decouple them by applying a convection boundary condition, Equation 2.9.

If there is a convective flow with specified flow velocity $\vec{v}$ within a computational domain, the heat transfer equation, Equation 2.5, gains an additional, so-called "thermal flow term" $\rho C_p \vec{v} \nabla T$. This term complicates the analytical solutions. However, numerical methods for solving Equation 2.5 can treat the thermal flow term with little additional effort and, moreover MOR-based solutions can be used as well [59].

If the flow velocity within a computational domain cannot be specified, as in the extreme case of bubble formation by microfluidic devices for example, there is no way of decoupling Equation 2.5 from Navier–Stokes equations. Hence, the solution of a spatially discretized coupled system becomes time consuming.

In the case of thermo-mechanical coupling, a good approximation is the solution of a thermal problem on the Lagrangian grid [60]. Then, it is possible to solve the thermal problem on its own and to use the temperature distribution to estimate additional stresses, which is due to thermal expansion.

The electro-thermal coupling is present through the heat generation rate Q, emerging mostly from the Joule effect, Equation 2.3. In general, in order to solve for the current in Equation 2.3, it is necessary to solve the Poisson equation, which (in the case of an isotropic resistive heater which is free from electrical charges) has the form:

$$\nabla \bullet j = \nabla \bullet (\sigma \nabla \varphi) = 0 \tag{2.11}$$

It is possible to solve the Poisson equation for the unknown electric potential field, provided that the geometry of the heater, its specific conductivity and the boundary conditions are specified [35]. Equation 2.11 depends only implicitly on time, due to possible changes in boundary values or changes in conductivity. This is because the speed of electron propagation is very high. We assume that Equation 2.11 holds at any given time. It should be noted that if the electrical current frequency reaches the megahartz range, i.e. when capacitance and inductive effects have to be accounted for, a higher level model based on solution of the electromagnetic Maxwell equations is required.

By inserting Equation 2.11 into Equation 2.3 the heat generation rate within an ohmic conductor changes to

$$Q = \frac{j^2}{\sigma} = \sigma \cdot (\nabla \varphi)^2 \tag{2.12}$$

By assuming homogeneous heat generation over a lumped resistor Equation 2.12 simplifies to the well known

$$Q = I^2 R = \frac{U^2}{R} \tag{2.13}$$

where I is the current passing through the lumped resistor, due to potential difference U over it, R is its resistivity and Q is the total heat generated within the resistor's volume. By replacing Q in Equation 2.5 with Equation 2.13, coupling to the Poisson equation is avoided. However, Equation 2.13 may not offer a good approximation in all cases. For example if resistors with complex geometries are present, homogeneous heat generation can not be assumed and in this case a consistent electro-thermal simulation is needed (that is, a simultaneous solution of Equation 2.12 and Equation 2.5), as shown in Figure 2.5.

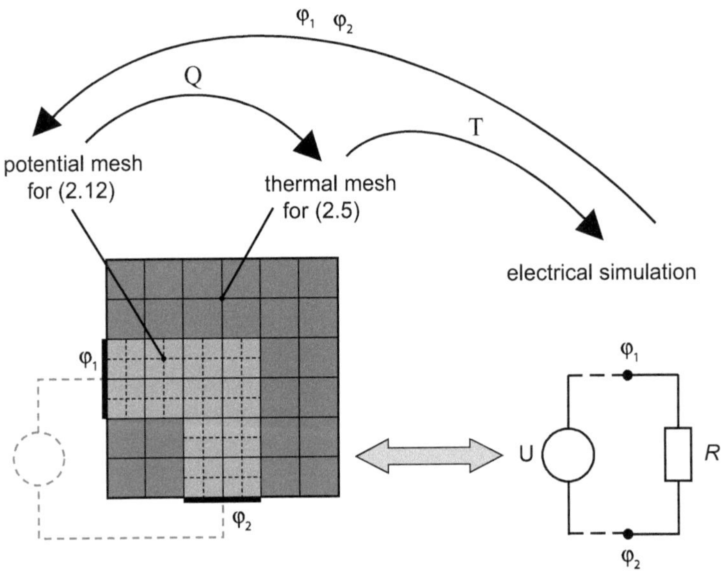

Fig. 2.5. Sequence of coupled electro-thermal simulation when coupling is at the device and at the system level.

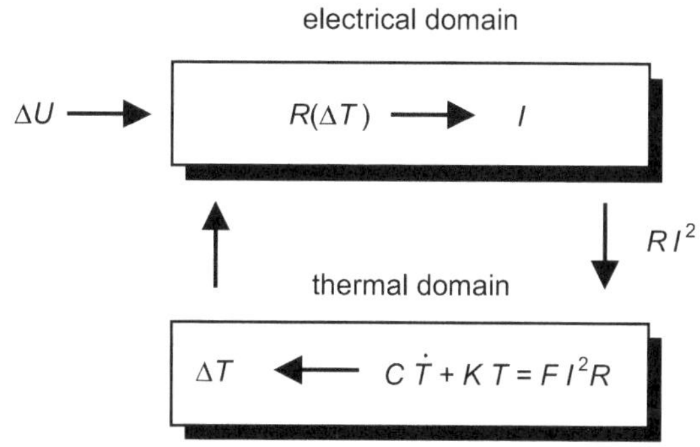

Fig. 2.6. Schematic of coupled electro-thermal simulation at system level for resistively heated microsystems. Homogeneous heat generation is assumed.

The electro-thermal coupling at the system level (assuming a homogeneous heat generation within a lumped heater) for the special case of resistively heated microdevices is schematically presented in Figure 2.6. Note that the coupling becomes more complicated in the case of ICs, i.e. when the temperature impact on semiconductor devices is taken into account [63].

In order to obtain a dynamic thermal system:

$$C\dot{T} + KT = F \cdot Q \tag{2.14}$$

shown in Figure 2.6 a spatial discretization of Equation 2.5 over the complete simulation domain (e.g. the whole chip) is necessary. Obviously, in order to effectively perform a system-level simulation, it is necessary to keep the dimension of Equation 2.14 as moderate as possible. This is exactly the goal of compact modeling described in Section 2.6.

2.5 Solving the Heat Transfer Partial Differential Equation

In the following as a transition between analytical and numerical solutions of Equation 2.5, we will use discretization of the computational domain in space, which necessarily leads to an ordinary differential equation system. One can start with a set of functions in the computational domain $f_i(x,y,z)$ for an infinite series expansion and then proceed systematically to determine the unknown coefficients $a_i(t)$ within the series, in order to solve the heat transfer equation. The separation of variables method gives:

$$T(x,y,z,t) = \sum_{i=0}^{\infty} a_i(t) f_i(x,y,z) \tag{2.15}$$

In general, if the expansion functions a_i and f_i exactly satisfy the BCs for the problem, we refer to the solution as analytical. An alternative is to split the domain into many small pieces, that is to discretize it in space. Then one can approximate the temperature field within an element with local shape functions and express the whole temperature field in a piece-wise fashion. In this case, we refer to a solution as numerical.

2.5.1 Linearization

When material properties in the heat transfer equation are constant, it is called linear. In this case, there are many mathematical benefits, for example one can use the superposition principle. Computationally it is definitely much easier to solve a linear partial differential equation. Almost all analytical solutions require the heat transfer equation to be linear. For numerical methods this is not required but it is no doubt advantageous to do so if possible. In general, however, the material properties are temperature dependent and Equation 2.5 has a form:

$$\nabla \bullet (\kappa(T)\nabla T) + Q - \rho C_p(T)\frac{\partial T}{\partial t} = 0 \tag{2.16}$$

It is always possible to perform linearization around an operating point (temperature) in order to convert a nonlinear heat transfer equation to a linear one. After obtaining a set of linear models around a chosen set of temperatures in this way, one can use a sort of weighting function, as done in [61], to extract a nonlinear compact thermal model.

In some special cases it is further possible to use certain transformations that linearize Equation 2.16. Under the assumptions that only the thermal conductivity κ is temperature dependent and that κ is not a function of space, the authors in [62] suggest using the Kirchoff transformation of the form:

$$\theta = T_s + \frac{1}{\kappa_s}\int_{T_s}^{T} \kappa(T)dT \tag{2.17}$$

with $\kappa_s = \kappa(T_s)$ and T_s the heat sink temperature, which fully linearizes the static part of Equation 2.16. By additionally defining a new time variable, τ:

$$k_s\tau = \int_0^t k(\theta)dt \tag{2.18}$$

where diffusivity $k_s = \kappa_s/\rho C_p$, the time-dependent nonlinear heat transfer equation, Equation 2.16, becomes fully linearized:

$$\nabla^2\theta + \frac{Q}{\kappa_s} - \frac{1}{k_s}\frac{\partial\theta}{\partial\tau} = 0 \tag{2.19}$$

2.5.2 Analytical Solutions

Most published analytical solutions assume a simple geometry of the computational domain. Examples are the thin rectangular heat source or a volume heat source on top of an infinite medium. The latter is often used for modeling a bipolar transistor (see [63] and the references there). Another example is a stacked layer structure that can be infinitely extended in length and with directions, used for modeling a power FET transistor [64]. The most general analytical solution so far is based on [62], in which a linearized heat conduction equation, Equation 2.19, is analytically solved in the rectangular (multilayer) thermal domain with arbitrarily distributed volume heat sources. It is, in principle, possible to divide a region with arbitrary geometry into rectangular thermal subvolumes and to apply the solution from [62]. However, in this case the problem of coupling these subvolumes must be solved as well.

Different analytical approaches use Green functions [65, 63], Fourier series [62, 66] or Fourier transforms [64].

As a conclusion to this subsection we want to state that although the analytical solutions require less computational effort and are preferred by microsystem designers in the above-mentioned special cases, it is unfortunately difficult to use them in the general case.

2.5.3 Numerical Methods

There are several methods associated with a mesh, which partitions an arbitrary computational domain into smaller units. These are the finite difference method (FDM), the finite volume method (FVM), the finite element method (FEM) and the boundary element method (BEM). They semi-discretize the heat transfer partial differential equation, Equation 2.5, and transform it into a system of ordinary differential equations, Equation 2.14. An overview of these methods and a discussion of similarities and differences between them can be found in [3]. For the heat transfer equation, Equation 2.5, a transmission line matrix (TLM) approach can be used as well [67, 68]. This approach represents a physical model of heat flow as a sequence of voltage (temperature) pulses traveling through a matrix network of transmission lines. This method requires a rectangular mesh and the creation of an *RC* network.

Lastly, each spatial discretization can be transformed into an equivalent thermal *RC* network. In this subsection we will briefly discuss the resemblance of thermal and electrical circuits.

We have mentioned the importance of creating a boundary condition independent (BCI) model. Let us briefly explain how a typical commercial solver generates such a model. The first step is the "pure" discretization of the computational domain without having specified the heat generation rate or boundary conditions. It results in two system matrices, C_{BCI} and K_{BCI}, which are stored in the software database. In the general case, these matrices depend on temperature and are not formed explicitly but rather are stored as a list of element matrices. Both are sparse and C_{BCI} is quite often lumped, which means that it is converted into a diagonal matrix. The dimension of C_{BCI} and K_{BCI} equals the number of introduced finite element nodes. We refer to these matrices as BCI, because it is possible to apply boundary conditions to a model described with C_{BCI} and K_{BCI} without repeated discretization.

After the Neumann and convection boundary conditions have been introduced, a system of ordinary differential equations is obtained as

$$C_{BCI}\dot{T}+K_{BCI}T+\sum K_{con}T=\sum f_{source}+\sum f_{Neumann}+\sum f_{con} \tag{2.20}$$

where the sum contains all the heat sources and boundary conditions. The volume heat sources and the Neumann boundary conditions contribute only to the load vector, while the convection boundary conditions contribute to both the load vector and the heat conductivity matrix. It is possible to add scaling factors to each heat source and boundary condition. Hence, Equation 2.20 can describe various external conditions without further need for any changes.

The use of Dirichlet boundary conditions, however, changes Equation 2.20 as follows:

$$C\dot{T}+KT=\sum f_{source}+\sum f_{Neumann}+\sum f_{con}+\sum f_{Dirichlet} \tag{2.21}$$

where the dimension of all vectors and matrices is reduced by the number of nodes to which the constant temperature is applied (Dirichlet nodes). Equation 2.21 is obtained from Equation 2.20 by crossing out columns and rows belonging to the Dirichlet nodes. C and K are obtained from C_{BCI} and K_{BCI} by crossing out those columns and rows which belong to the Dirichlet nodes. However, the use of Dirichlet boundary conditions makes Equation 2.21 less flexible, as it is impossible to replace them with different boundary conditions once they have been applied. In Equation 2.20, one can change BCs by simply using zero scaling factors. Fortunately, one can always replace the Dirichlet boundary conditions with the convection BCs if a very high value for the film coefficient is used.

Equation 2.20 and Equation 2.21 are equivalent to Equation 2.14 and can be directly plugged into the electric simulator as an equivalent *RC*-network. In order to highlight a resemblance of thermal and electrical circuits, let us consider the parallel *RC* circuit shown in Figure 2.7.

It is described by:

$$C\begin{bmatrix}1 & -1\\ -1 & 1\end{bmatrix}\begin{bmatrix}\dot{V}_1\\ \dot{V}_2\end{bmatrix}+\frac{1}{R}\begin{bmatrix}1 & -1\\ -1 & 1\end{bmatrix}\begin{bmatrix}V_1\\ V_2\end{bmatrix}=\begin{bmatrix}I_1\\ I_2\end{bmatrix} \tag{2.22}$$

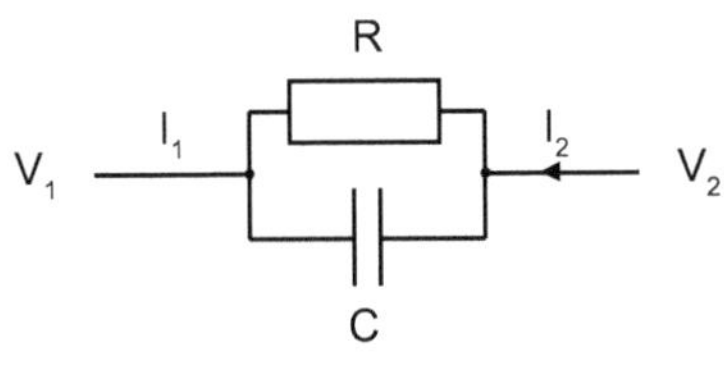

Fig. 2.7. A parallel *RC* circuit.

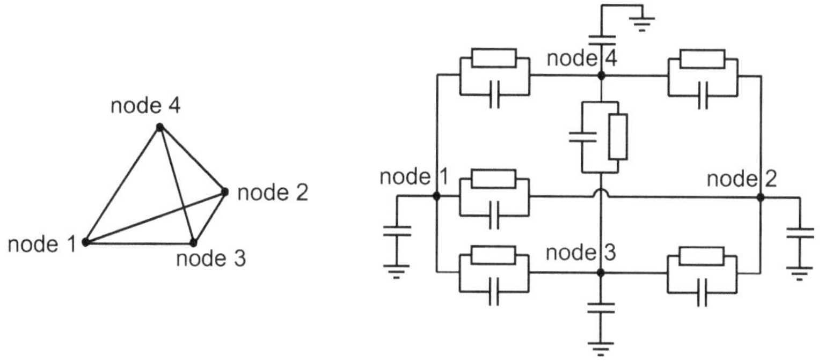

Fig. 2.8. Thermal impedance network for a tetrahedral element for use away from the boundaries, adapted from [69].

When comparing Equation 2.22 with Equation 2.14 it is easy to see that the heat capacity matrix can be realized by capacitor elements, the heat conductivity matrix by resistor elements and the heat source vector (matrix) by current sources. The equivalent thermal networks were derived for different types of finite elements [69]. Figure 2.8 shows a conductive thermal network for a tetrahedral element with a convective boundary.

It should be noted that the transformation of Equation 2.14 into an equivalent thermal *RC* network is exact, i.e. no approximation (in the sense of compacting the circuit) is made so far.

2.6 Dynamic Compact Thermal Modeling

As already mentioned, it is the dimension of Equation 2.14 which makes the simulation time consuming. For a variety of MEMS devices the order of the resulting thermal ODE system exceeds 100000. Hence, it is prohibitive to use these models during system-level simulation. Instead, accurate dynamic compact thermal models are required. The methods for constructing dynymic compact thermal models can be divided into three general groups: non-automatic methods such as different *RC* ladder network approaches, semi-automatic methods such as modal approaches and MOR methods which can be made fully automatic. Through this section we will describe the main properties of the first two groups. MOR methods will be the topic of Chapter 3.

2.6.1 *RC* Ladder Approach

A large number of dynamic compact thermal modeling approaches are based on fitting an *RC* ladder network to the observed (measured or numerically computed) system response, using a suitable optimization technique. In this case, however, the *RC* ladder network is based on an attempt to lump a distributed thermal domain. For 1D heat conduction, if we subdivide the domain into a large number of small slabs whose thicknesses L go to zero, we get a so-called Cauer ladder network (see Figure

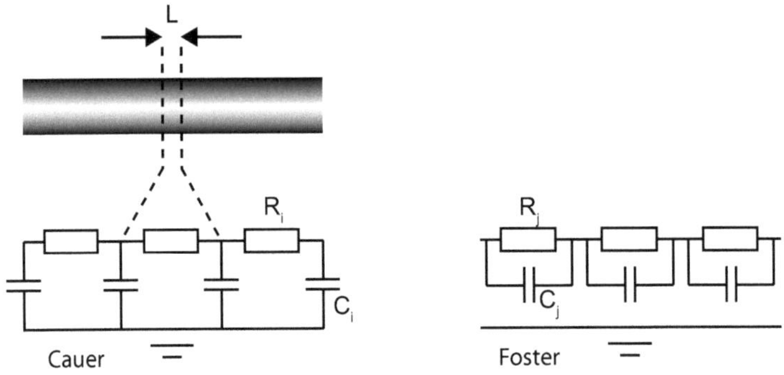

Fig. 2.9. Cauer *RC* ladder network (left) and Foster network (right).

2.9 left) with grounded capacitors and floating resistors. This can be transformed into a Foster network (Figure 2.9 right) via standard circuit transformation algorithms. While the Cauer network appears to be a small version of the large equivalent circuit described in the previous section (that is a fine mesh replaced by a coarse one), the Foster network has no physical meaning. In both cases, the question is how to set a proper number of *RC* pairs or, in the case of 2D and 3D thermal conduction, how to choose a proper network structure at all.

In [71] the authors suggest a robust method based on computing a time-constant spectrum function connected to a Foster network. A Foster network representation brings along an infinite number of time-constants (each R_iC_i pair represents a time constant $\tau_i = C_i \cdot R_i$) and so defines a continuous τ spectrum. The step response of a single R_iC_i stage is $R_i(1-\exp(-t/\tau_i))$. Hence, the unit-step response of the Foster network can be constructed as a sum of these exponential terms:

$$T(t) = \sum_i R_i(1-\exp(-t/\tau_i)) \tag{2.23}$$

After replacing R_i with a continuous spectrum, the above sum can be replaced by the integral over the whole τ range:

$$T(t) = \int_{-\infty}^{\infty} R(\xi)(1-\exp(-t/\exp(\xi)))d\xi \tag{2.24}$$

where $R(\xi)$ is the time-constant spectrum function defined on the $z = \ln(t)$ and $\xi = \ln(\tau)$ logarithmic time axes. Lastly, the relationship between $T(t)$ and spectrum can be expressed by convolution as

$$\frac{d}{dz}T(z) = R(z) \otimes w(z) \tag{2.25}$$

where $w(z) = \exp(z-\exp(z))$. Hence, from the measured step response $T(t)$ we can determine a suitable number of *RC* ladder elements in the Foster circuit required to represent a (preferably multilayer) system prior to fitting. As the Foster network has no physical meaning, it has become an engineering practice to convert it back to Cauer form. This method also works for the 2D and 3D heat conduction, but is unfortunately applicable to only a single conduction path, i.e. to single-input–single-output (SISO) problems.

For the multiple-input–multiple-output (MIMO) system, not only the number but also the structure of a compact *RC* ladder network must be specified. Recently, the authors in [54] suggested the application of evolutionary algorithms for setting the best topology of the compact model.

A special challenge in dynamic compact thermal modeling is to construct a boundary condition independent compact model, which would be reusable for different surroundings. This means that if, for example a chip producer does not know the conditions under which the chip will be used, the compact thermal model must allow an engineer to research how environmental changes influence the chip temperature. The two European projects, DELHPI and PROFIT, have addressed the need to produce an accurate and boundary condition independent compact thermal models of a chip [72–76], which would simplify the chip cooling simulation of coupled thermal and fluidic domain. The chip benchmarks representing boundary condition independent requirements are described in [77]. Related discussions can be also found in [78 and 79]. The goal of the PROFIT project was to extend the methodology to transient compact thermal models by using methods from [71 and 80]. The current solutions from both projects are mainly based on data fitting for an *a priori* chosen resistor network.

In spite of the large number of related methods suggested in recent years (an extensive review can be found in [81]), *RC* network extraction remains a non-automatic approach, which requires a designer to choose the correct number and position of *RC* ladders without strict guidelines, and to perform a time-consuming parametrization.

2.6.2 Modal Approaches

It is well-known from structural mechanics that an elastic string vibrating under the action of arbitrary external forces, can be viewed as a linear combination of different vibrating modes, each one corresponding to a resonance of the string. Although their number is infinite, the actual response can be approximated quite well by considering only the first few harmonics. This is how the elastic string, which is a distributed system, can be compacted into a lumped system with only a few degrees of freedom. In terms of the partial differential equation governing string vibration this means that one has to compute the eigenvalues, which are the vibration frequencies, and the eigenfunctions, which are the spatial forms of vibrations for each mode. Out of the first few eigenfunctions a compact model can be constructed.

In an analog manner, the temperature can be expressed as a series expansion around a set of "dominant" eigenfunctions U_j:

$$T(r,t) \cong \sum_{j=1}^{k} U_j(r) \cdot V_j(t) \tag{2.26}$$

where V_j are jet unknown expansion coefficients. By substituting Equation 2.26 into Equation 2.5, multiplying it by V_i and integrating, it is possible to get a reduced model [81] of the form:

$$C^* dV(t)/(dt) + K^* V(t) = s(t) \tag{2.27}$$

where

$$C_{ij}^* = \int \rho C_p U_i U_j dr,\ K_{ij}^* = \int U_i \nabla(\kappa \nabla U_j) dr,\ s_i = \int U_i Q dr,\ i,j \in [1,k].$$

Unfortunately, an analytical solution of the eigenvalues problem for the heat conduction partial differential equation, Equation 2.5, is available only for some simple geometries [82]. Hence, a first approximation is to perform spatial discretization and to compute the eigenvalues and corresponding eigenvectors of the system matrix of Equation 2.14. To do this, we need to reduce Equation 2.14 to a single matrix representation:

$$A\dot{T} = T + Bu(t) \tag{2.28}$$

where $A = -K^{-1}C$ and $B = -K^{-1}FQ$. The number of eigenvectors of A equals the number n of finite element nodes. It is possible to reduce Equation 2.28 by projection, where a projection matrix $V_{modal} \in C^{n \times r}$ is composed of r eigenvectors of A. Now the n-dimensional Equation 2.28 can be projected onto r-dimensional subspace as follows:

$$V_{modal}^T \cdot A \cdot V_{modal} \cdot \dot{z} = V_{modal}^T \cdot V_{modal} \cdot z + V_{modal}^T \cdot Bu(t) \tag{2.29}$$

where z is a generalized variable vector. The question which remains is how to choose the "dominant" modes (eigenvectors). For the defined thermal output $y(t) = E^T \cdot T$ (E is either a vector or matrix), one can observe the transfer function $G(s) = E^T(sI - A)^{-1}B$, and choose those poles (and associated eigenvalues) which are around the region of frequencies of interest. Nevertheless, the choice of important modes still requires designer action, which makes a modal approach manual.

2.6.3 Model Order Reduction

The only group of dynamic compact thermal modeling methods which can be made fully automatic, i.e. with only minimal intervention by the designer, are the mathematical MOR methods. Hence, among microelectronic and MEMS designers they are becoming increasingly popular [83–89].

The ideas of mathematical MOR have been developed in control theory and are applicable to first-order linear ODE systems, such as Equation 2.14. The development of model reduction of nonlinear systems is still in its early stages [90, 46].

According to control theory, Equation 2.14 should be written in the right-hand side state-space formulation as

$$\begin{aligned} \dot{T} &= AT + Bu(t) \\ y(t) &= E^T \cdot T \end{aligned} \tag{2.30}$$

where the system is treated as a black box, i.e. the internal state vector of temperatures $T \in R^n$ is not directly accessible to an external observer. The controller can influence the system state through the input functions specified by the vector $u(t) \in R^m$ and distributed to the internal nodes in accordance to the input matrix $B \in R^{n \times m}$. As the number of inputs is typically small $m \ll n$, the matrix B has a small number of columns. Furthermore, the observer is interested in only a few outputs, specified by the vector $y \in R^p$. The required outputs are selected from the complete state vector via output matrix $E \in R^{n \times p}$. Hence, a high-dimensional ODE system, governed by a small number of external inputs, has to be solved in order to determine a small number of relevant outputs.

Let us also mention here the all-important transfer function of the system from Equation 2.30, defined as

$$G(s) = E^T (sI - A)^{-1} B \tag{2.31}$$

where s is the Laplace variable and I is a unity matrix of dimension n. Although $G(s)$ is a relatively small matrix with p rows and m columns, its computation requires the inverse of a large-scale system matrix A.

While searching for a possibility to reduce the number of state variables, i.e. ODEs in Equation 2.30, let us transform the state vector T using a transformation matrix $V_n \in R^{n \times n}$ as follows:

$$T = V_n \cdot z \tag{2.32}$$

Here, z is the new state vector in terms of generalized coordinates expressed by the transformation matrix. It is important to understand that if T has the spatial and physical meaning of a vector of temperatures belonging to FE nodes, z has none of either. Rather, it is a vector of multiplication factors for the "global shape functions" given by the columns of matrix V_n.

To make this more clear, let us take a closer look at discretization methods where the temperature over the whole domain is approximated as a piece-wise linear combination over the elements:

$$T(r) = \sum_{\Omega_1}^{i=1,n} c_i^1 N_i^1(r) + \ldots + \sum_{\Omega_e}^{i=1,n} c_i^e N_i^e(r) \tag{2.33}$$

where n is the number of free coefficients c_i in each element (these can be the node temperatures for example) and e is the total number of elements. A property of each local shape function N_i is that it is one at the ith node and zero outside the finite element. Within the element it can linearly decay between one and zero. The only way

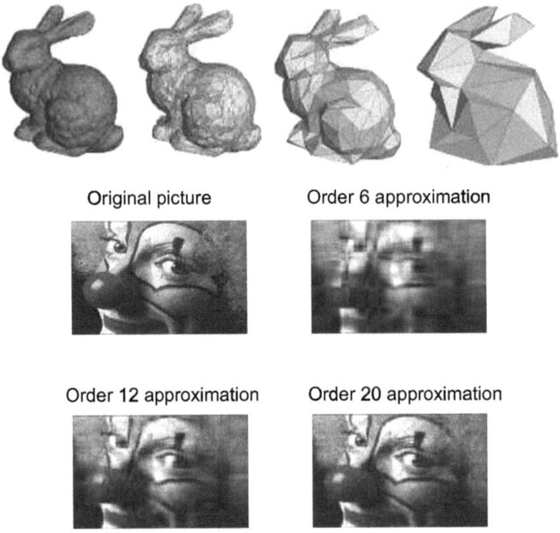

Fig. 2.10. Order reduction by coarsening the mesh (top, by courtesy of H. Hoppe) and by truncating the generalized coordinates (bottom, by courtesy of A. Antoulas).

to directly reduce a system based on Equation 2.33 is to coarsen the mesh, i.e. to choose shape functions which cover several elements. This, however, may result in a significant loss of precision (see Figure 2.10 top).

On the other hand, each column of matrix V_n in Equation 2.32 can be seen as a linear combination of the local shape functions, i.e. as a global shape function over the whole heat transfer domain. This gives hope that it may be possible to truncate some of the generalized coordinates z and therefore reduce the dimension of the system Equation 2.32 without losing much accuracy (see Figure 2.10 bottom).

Both ways of compressing information, which are shown in Figure 2.10, have their advantages and disadvantages. By coarsening the mesh, one preserves the physical nodes, but loses precision. By performing a mathematical MOR, one loses the physical nodes, but preserves high accuracy (as will be explained below). Note that those two methods do not exclude each other, but can be combined in order to maximally increase the efficiency.

Let us emphasize that after the transformation, Equation 2.30 changes to:

$$\begin{aligned} \dot{z} &= W_n^T A V_n \cdot z + W_n^T B u(t) \\ y(t) &= E^T V_n \cdot z \end{aligned} \tag{2.34}$$

where $W_n, V_n \in R^{n \times n}$ are biorthogonal, which means that $W_n^T \cdot V_n = I$. For the external observer Equation 2.34 behaves exactly the same way as Equation 2.30, i.e. one can easily prove that the transfer function, Equation 2.31, does not change.

As already mentioned, model reduction is based on the idea that one can find such a transformation when one can accurately represent the state vector with just a

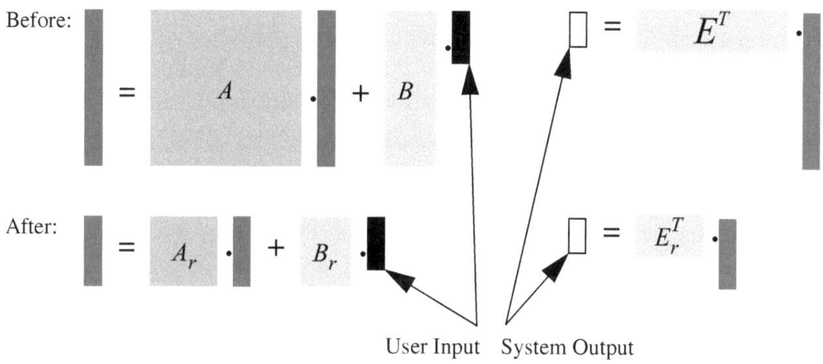

Fig. 2.11. Schematics of the system before and after model reduction, adapted from [6].

few generalized coordinates. In other words, a transformation exists when one can truncate most of the generalized coordinates, that is to approximate:

$$T = \begin{bmatrix} V_r \; V_{n-r} \end{bmatrix} \cdot \begin{bmatrix} z_1 \\ z_2 \end{bmatrix} \tag{2.35}$$

through

$$T = V_r \cdot z_1 + \varepsilon \tag{2.36}$$

with the error vector ε being small. This, on the other hand, changes Equation 2.34 to a low-dimensional system:

$$\begin{aligned} \dot{z}_1 &= A_r z_1 + B_r u(t) \\ y_r(t) &= E_r^T \cdot z_1 \end{aligned} \tag{2.37}$$

with $A_r = W_r^T A V_r$, $B_r = W_r^T B$, $E_r = V_r^T E$ and $W_r^T \cdot V_r = I$. The number of inputs and outputs in Equation 2.37 is the same as in Equation 2.30, whereas the number of equations (dimension of the state vector) is smaller. This transformation is schematically shown in Figure 2.11.

The Equation 2.36 can be also seen as a projection of an n-dimensional state vector to a r-dimensional subspace $r \ll n$, defined by the columns of matrix V_r (global shape functions). Equation 2.37 is a projection of the whole system in Equation 2.30. The output of the reduced system is, however, not the same as that of Equation 2.30 or Equation 2.34, since we have introduced the truncation error ε. The goal of MOR is to minimize this error either in the time domain $\min\|y(t) - y_r(t)\|$ or in the Laplace domain $\min\|G(s) - G_r(s)\|$, where the transfer function of the reduced system is defined as

$$G_r(s) = E_r^T (sI - A_r)^{-1} B_r \tag{2.38}$$

There are several model reduction methods for linear ODE systems which produce small ε. They take the system matrices A and B of the linear system as input and perform linear algebra manipulations with them in a different manner to construct W_r and V_r. Let us emphasize that, unlike the non-automatic *RC*-ladder approach, there is no explicit minimization procedure for ε and in a way, one obtains the best topology of a reduced system simultaneously with its system matrices. These methods and their most important properties will be discussed in Chapter 3.

Finally, let us get back to the physical sense of the internal nodes in the reduced models. The goal is to approximate the original system, which means, the smaller the error, the more the physical sense. If we think of *RC* parameters as the amplitudes for global functions, then any network, including the Foster *RC*-network, has a valid physical sense for thermal modeling.

2.7 Summary

In this chapter we have highlighted the importance of heat transfer modeling, which is presently a central part of the electro-thermal simulation of microsystems. We have discussed in which cases it is possible to decouple the heat transfer PDE from other physical domains, in order to be able to solve it separately. Unfortunately, analytical solutions are only available for simple geometries, whereas either approximations or numerical methods must be used for complex geometries.

However, the numerical solution of the heat transfer PDE, e.g. via finite elements, is often impractical or even prohibitive if we want to simulate the whole system with a large number of interconnected devices and driving circuitry. Again, the number of resulting ordinary differential equations for a single device easily exceeds 100000. Although, a domain decomposition technique combined with modern hardware allows solution of systems with hundreds of millions ODEs, the large demand for computational resources inhibits system-level simulation. Hence, a reduction of the number of unknowns in the system under study, known as dynamic compact thermal modeling, has become a standard for microsystem simulation.

We have presented the conventional, non-automatic approaches (*RC*-ladder network), semi-automatic approaches (modal approximation), and the increasingly popular MOR approaches, which can be made fully automatic. However, we emphasize dynamic compact thermal modeling via MOR, which follows spatial discretization, because it presently offers an accurate and effective solution for electro-thermal modeling of microsystems and enables automatic system-level modeling.

A schematics of the recommended research paths described in this chapter is shown in Figure 2.12.

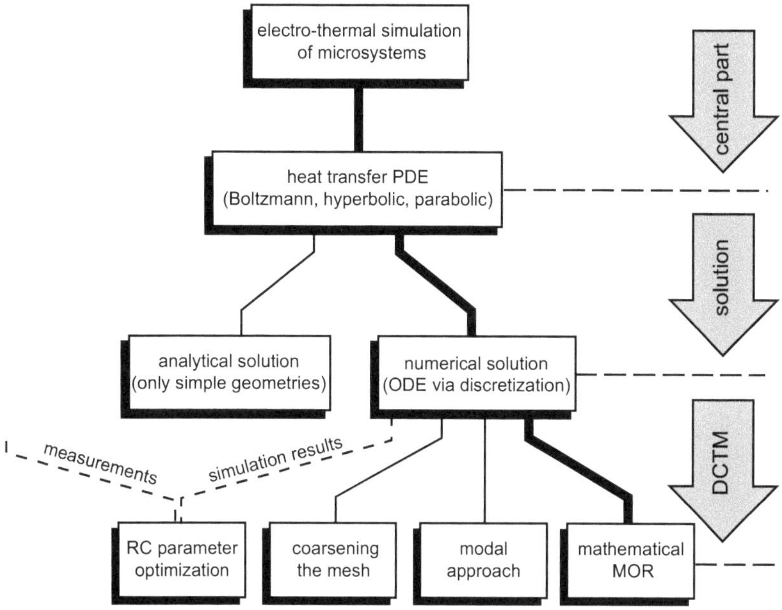

Fig. 2.12. Schematics of recommended (bold) research paths in electro-thermal modeling of microsystems.

3

Overview of Linear Model Order Reduction

In the previous chapter we highlighted the importance of model order reduction (MOR) techniques for dynamic compact thermal modeling of microsystems. The goal of this chapter is to describe a number of relevant MOR methods for linear systems. Let us emphasize that for MEMS case studies in this book, we assume material thermal properties κ and C_p to be temperature independent, which means that spatial discretization results in a linear ODE system as in Equation 2.14. State of the art MOR for nonlinear systems will be presented in Chapter 6.

Section 3.1 introduces the mathematically mature control theory methods, such as balanced truncation approximation (BTA), singular perturbation approximation (SPA) and Hankel norm approximation (HNA). Section 3.2 introduces two Krylov subspace methods, the Lanczos algorithm and the Arnoldi process and compares their performances. It also explains the terms Padé approximants and moment matching in more details. In Section 3.3 the Guyan reduction, which has become a part of commercial FE simulators, like ANSYS and ABACUS, is presented. Section 3.4 summarizes the most important characteristics of all presented methods.

3.1 Control Theory Methods

We start from the stable linear dynamic state space system of the form:

$$\begin{aligned} \dot{x}(t) &= Ax(t) + Bu(t) \\ y(t) &= E^T \cdot x(t) \end{aligned} \tag{3.1}$$

Here t is the time variable, $x(t) \in R^n$ is a state vector, $u(t) \in R^m$ the input excitation vector, and $y(t) \in R^p$ the output measurement vector. $A \in R^{n \times n}$ is the system matrix, $B \in R^{n \times m}$ and $E \in R^{n \times p}$ are input and output distribution arrays, respectively. n is the state space dimension, m and p are the number of inputs and outputs, respectively. As already mentioned in the previous chapter, in most practical cases we can assume that m and p are much smaller than n.

Control theory offers a number of well-established tools for the automatic model reduction of stable linear systems [93]. Before describing different methods, let us introduce two important characteristics of the system in Equation 3.1 called controllability and observability. A system is said to be controllable if for any initial state $x(t_0) = x_0$ and any final state x_1, there exists an input $u(t)$ that transfers x_0 to x_1 in a finite time. It is observable if it is possible to calculate any unknown initial state $x(t_0) = x_0$ uniquely from the given input $u(t)$ and the measured output $y(t)$ over a finite time interval. In the following we will assume the system in Equation 3.1 to be a so called minimal realization, i.e. there are no states which are neither controllable nor observable at all. Note that the system resulting from spatial discretization of computational domains (like the MEMS case studies in this book) are always minimal realizations. In terms of control theory, controllability and observability are described by means of the gramians, P and Q, defined as:

$$P = \int_0^{\infty} e^{At} BB^T e^{A^T t} dt$$
$$Q = \int_0^{\infty} e^{A^T t} E^T E e^{At} dt \tag{3.2}$$

It can be shown [95] that the criteria for controllability and observability are equivalent to the condition that P and Q are both regular (i.e. have full rank). If, in addition, all eigenvalues of A have negative real parts (i.e. the system is stable), then the gramians can be computed as the unique and positive definite (symmetric real matrix A is positive definite if $x^T Ax > 0, \forall x \neq 0$) solutions of two Lyapunov equations related to the system in Equation 3.1:

$$AP + PA^T = -BB^T$$
$$A^T Q + QA = -E^T E \tag{3.3}$$

Numerically stable and direct methods to solve Lyapunov equations and find the controllability and observability gramians can be found in [9].

3.1.1 Balanced Truncation Approximation

The idea behind model reduction via balanced truncation is to obtain a reduced-order model by finding and deleting those states that are simultaneously least controllable and observable. Moore [94] suggested such a similarity transformation, which balances the system, i.e. sets both gramians equal and diagonal:

$$\hat{P} = \hat{Q} = diag(\sigma_1, \ldots, \sigma_n) \tag{3.4}$$

with the diagonal entries in descending order $\sigma_1 \geq \sigma_2 \geq \ldots \geq \sigma_n$. The parameters σ_i are called Hankel singular values (HSV) and are transformation invariant. They are a property of the system and depend only on input–output behavior. Hankel singular values can be computed as the square roots of eigenvalues λ_i of the product of P and Q:

$$\sigma_i = \sqrt{\lambda_i(P \cdot Q)}, i = 1, \ldots, n \tag{3.5}$$

It can be shown that the Hankel singular values reflect the contributions of different entries of the state vector to system responses [95, 10]. Hence, to reduce the order of the balanced system ($\tilde{A}, \tilde{B}, \tilde{E}$), it is sufficient to remove (truncate) those state variables (and corresponding blocks in $\tilde{A}, \tilde{B}$ and $\tilde{E}$) related to the smallest Hankel singular values. Balancing and truncation can be done simultaneously by applying a projection to the original non-balanced system, using, for example, the square root algorithm (Algorithm 3.1) [9, 96].

Algorithm 3.1 *The square root algorithm for calculating BTA of order r*

Inputs: A , B , E

Outputs: $\tilde{A}_r$, $\tilde{B}_r$, $\tilde{E}_r$ *the matrices of the reduced, internally balanced system of order r,* W_r *and* V_r *– mutually orthogonal projection matrices such that* $\tilde{A}_r = W_r^T A V_r$, $\tilde{B}_r = W_r^T B$ *and* $\tilde{E}_r = V_r^T E$

1. Compute the controllability gramian P and the observability gramian Q by solving the Ljapunov equations.

2. Find the Cholesky factorization of the solutions P and Q:

$$P = L_c L_c^T, \quad Q = L_o L_o^T$$

3. Calculate the singular value decomposition of the matrix $L_o^T L_c$:

$$L_o^T L_c = U_1 \Sigma U_2$$

where $\Sigma \in R^{n \times n}$ *is a diagonal matrix of Hankel singular values in descending order.*

4. Form the projection matrices $W_r \in R^{n \times r}$ *and* $V_r \in R^{n \times r}$ *as*

$$V_r = L_c U_{2_r} \Sigma_r^{-\frac{1}{2}} \text{ and } W_r = L_o U_{1_r} \Sigma_r^{-\frac{1}{2}}$$

where U_{1_r} *and* U_{2_r} *are the first r columns of matrices* U_1 *and* U_2, *and*

$$\Sigma_r^{-\frac{1}{2}} = diag\left(\frac{1}{\sqrt{\sigma_1}}, \ldots, \frac{1}{\sqrt{\sigma_r}}\right)$$

5. Apply the projection to the system (3.1) to find the order r truncated balanced realization as

$$\tilde{A}_r = W_r^T A V_r \ , \ \tilde{B}_r = W_r^T B \ , \ \tilde{E}_r = V_r^T E$$

The controllability and observability gramians of the reduced system ($\tilde{A}_r, \tilde{B}_r, \tilde{E}_r$) of order r are diagonal and equal, $\tilde{P}_r = \tilde{Q}_r = diag(\sigma_1, \ldots, \sigma_r)$. Some alternatives to Algorithm 3.1 can be found in [9].

The most important characteristic of Algorithm 3.1 is that it provides a global error bound between the transfer functions of the original and the reduced systems $G(s)$ and $G_r^{BTA}(s)$:

$$\left\| G(s) - G_r^{BTA}(s) \right\|_\infty \leq 2(\sigma_{r+1} + \ldots + \sigma_n) \tag{3.6}$$

where the infinity norm $\|.\|_\infty$ denotes the largest magnitude of the difference in transfer functions, and G(s) and $G_r^{BTA}(s)$ are defined as in Equation 2.31 and Equation 2.38, respectively. The proof of Equation 3.6 can be found in [10]. The discussion of different error norms is given in [93].

The practical consequence of Equation 3.6 is that the success of model reduction depends only on the decay rate of the Hankel values. Figure 3.1 shows the behavior of Hankel values for a sample thermal MEMS model of dimension 1000. In the figure, only the first 50 Hankel singular values are shown and it is worth noting the logarithmic scale for Hankel singular values. One can see, that Hankel singular values decay rapidly and one needs about 15 states corresponding to the highest Hankel singular values to accurately capture the dynamics of the system according to Equation 3.6.

Hence, the Hankel singular values computed for a particular application give us a complete answer as to what extent we can reduce the original system. Furthermore, MOR becomes completely automatic, as a user just sets an error bound and then, by means of Equation 3.6, the algorithm finds the smallest possible dimension of the reduced system, which satisfies that bound. Alternatively, a user specifies the

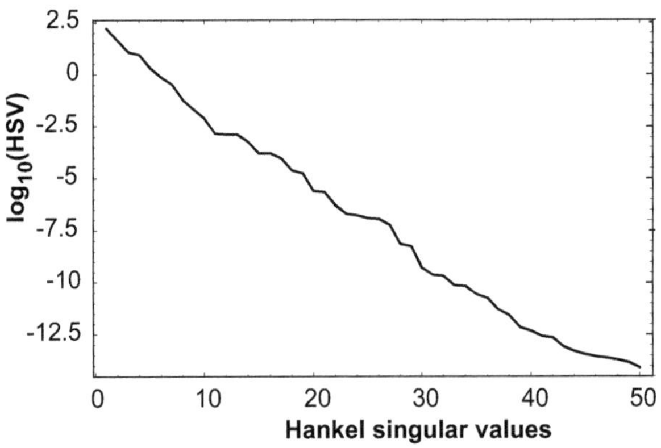

Fig. 3.1. Decay of Hankel singular values for the sample thermal MEMS device.

dimension of the reduced system and the algorithm estimates the error bound for the reduced system.

3.1.2 Singular Perturbation Approximation

The fact that the balanced truncation approximation generally incurs an approximation error in the low-frequency region is undesirable in some practical applications. Hence, an algorithm which produces zero error at zero frequency, called the singular perturbation technique, is obtained as follows. After decomposing the system matrices and the state vector of Equation 3.1 as

$$A = \begin{bmatrix} A_{11} & A_{12} \\ A_{21} & A_{22} \end{bmatrix}, \quad B = \begin{bmatrix} B_1 \\ B_2 \end{bmatrix}, \quad E = \begin{bmatrix} E_1 & E_2 \end{bmatrix}, \quad x = \begin{bmatrix} x_1 \\ x_2 \end{bmatrix} \tag{3.7}$$

with $A_{11} \in R^{r \times r}$, and assuming that A_{22} is invertible, the reduced system is defined by following truncation [9]:

$$\begin{aligned} \dot{z} &= (A_{11} - A_{12}A_{22}^{-1}A_{21})z + (B_1 - A_{12}A_{22}^{-1}B_2)u \\ y_r &= (C_1 - C_2A_{22}^{-1}A_{21})z - C_2A_{22}^{-1}B_2u \end{aligned} \tag{3.8}$$

An important property of Equation 3.8 is that the steady-state gain matches that of the original system, that is:

$$G_r^{SPA}(0) = G(0) \tag{3.9}$$

It turns out that if we first balance the system, and then truncate it as in Equation 3.8, the reduced system also fulfills the error bound given by Equation 3.6 (see [9] for details):

$$\left\| G(s) - G_r^{SPA}(s) \right\|_\infty \le 2(\sigma_{r+1} + \ldots + \sigma_n) \tag{3.10}$$

3.1.3 Hankel Norm Approximation

The Hankel norm $\|.\|_H$ of $G(s)$ is defined as the maximal Hankel singular value of the system in Equation 3.1:

$$\|G(s)\|_H = \sqrt{\lambda_{max}(P \cdot Q)} = \sigma_{max} \tag{3.11}$$

The optimal Hankel norm approximation problem is the problem of finding an approximation $G_r^{HNA}(s)$ of degree $r<n$ such that the Hankel norm of the error $\left\| G(s) - G_r^{HNA}(s) \right\|_H$ is minimized.

The lower bound for the above norm is given as follows [9]:

$$\sigma_{r+1} \le \left\| G(s) - G_r^{HNA}(s) \right\|_H \tag{3.12}$$

Table 3.1. Advantages and disadvantages of control theory methods

Method	Advantages	Disadvantages
BTA	simple implementation, short computational times	does not preserve steady state
SPA	approximates steady state	extra computational time
HNA	specifies additionally a lower error bound for the transfer function	complex implementation and no preservation of steady state

Equation 3.12 holds for any $G_r(s)$ with exactly r stable poles. Since the infinity norm is never smaller than the Hankel norm [10], Equation 3.12 also means:

$$\sigma_{r+1} \leq \left\| G(s) - G_r^{HNA}(s) \right\|_\infty \tag{3.13}$$

An algorithm for the construction of an optimal Hankel norm approximation can be found in [9]. An optimal Hankel norm approximation $G_r^{HNA}(s)$ of order r fulfills the following upper bound:

$$\left\| G(s) - G^{HNA}{}_r(s) \right\|_\infty \leq (\sigma_{r+1} + \ldots + \sigma_n) \tag{3.14}$$

which is half the bound for the balanced truncation case.

3.1.4 Comparison of Methods

The Hankel norm approximation is considered by mathematicians to be the optimal method in terms of additionally specifying the lower bound, Equation 3.13, for the infinity error norm [92]. Nevertheless, the most frequently used method is balanced truncation approximation, due to the simplicity of implementation and shorter computational time. A comparison of computational times for all three methods is given in [13]. It can be shown that all the methods preserve the stability of the original system in the reduced system. In Table 3.1 we summarize their advantages and disadvantages.

Apart from the three methods described, control theory deduces a number of other reduction algorithms that are suitable for linear systems. However, none of them offers an error estimate. A good review of available methods can be found in [95] or [98].

3.2 Krylov Subspace Methods

Similar to control theory algorithms, Krylov subspace techniques require a single matrix representation of a first-order ODE system. In this case, however, it is more convenient to choose a left-hand side representation of the form:

$$\begin{aligned} A\dot{x}(t) &= x(t) + bu(t) \\ y(t) &= e^T \cdot x(t) \end{aligned} \tag{3.15}$$

with the same syntax as in Equation 3.1. Note that in this section we consider only the single-input–single-output (SISO) case, i.e. $m = p = 1$ (for SISO we use the lower case letters b and e for the input and output arrays), which does not effect the generality of the methods. The SISO transfer function $G(s)$ is a scalar-valued rational function defined as

$$G(s) = -e^T(I - sA)^{-1}b \tag{3.16}$$

The Taylor series expansion of $G(s)$ about $s_0 = 0$ is given by

$$G(s) = -e^T(I + sA + s^2A^2 + \ldots)b = \sum_{i=0}^{\infty} m_i s^i \tag{3.17}$$

where $m_i = -e^T A^i b$ for $i = 0, 1, 2, \ldots$, are called moments about s_0.

Krylov subspace methods for MOR correspond to choosing $G_r(s)$ as a Padé or Padé-type approximant of $G(s)$. The rth Padé approximant of the transfer function $G(s)$ about the expansion point s_0, is a rational function:

$$G_r(s) = \frac{P_{r-1}(s)}{Q_r(s)} = \frac{a_{r-1}s^{r-1} + \ldots + a_1 s + a_0}{b_r s^r + b_{r-1}s^{r-1} + \ldots + b_1 s + 1} \tag{3.18}$$

whose Taylor series around s_0 agrees with the Taylor series of $G(s)$ in the first $2r$ terms (moments), i. e. for $s_0 = 0$:

$$G(s) = G_r(s) + O(s^{2r}) \tag{3.19}$$

where

$$\lim_{s \to s_0} O(s^{2r}) = 0 \tag{3.20}$$

We will also consider reduced order modeling based on functions $G_r(s)$ for which less than $2r$ moments match. In this case $G_r(s)$ is called a Padé-type approximant. We refer the reader to [99] for an overview of Padé and Padé-type approximants.

Equation 3.19 presents $2r$ conditions for the $2r$ degrees of freedom (coefficients) of $G_r(s)$. Note that coefficient b_0 in Equation 3.18 is set to one, which eliminates an arbitrary multiplicative factor in $G_r(s)$. The coefficients a_i and b_i of polynomials $P_{r-1}(s)$ and $Q_r(s)$ can be computed as follows. Multiplying by $Q_r(s)$ on both sides of Equation 3.19 yields

$$G(s)Q_r(s) = P_{r-1}(s) + O(s^{2r})Q_r(s) \tag{3.21}$$

If we develop $G(s)$ up to the moment m_{2r}, we get:

$$(m_0+m_1s+\ldots+m_rs^r+\ldots+m_{2r}s^{2r})(b_rs^r+b_{r-1}s^{r-1}+\ldots+b_1s+1)$$
$$=a_0+a_1s+\ldots+a_{r-1}s^{r-1}+O(s^{2r})(b_rs^r+b_{r-1}s^{r-1}+\ldots+b_1s+1)$$

which further yields:

$$\begin{aligned}
&m_0(b_rs^r+b_{r-1}s^{r-1}+\ldots+b_1s+1)\\
&+m_1s(b_rs^r+b_{r-1}s^{r-1}+\ldots+b_1s+1)+\ldots\\
&+m_{r-1}s^{r-1}(b_rs^r+b_{r-1}s^{r-1}+\ldots+b_1s+1)\\
&+m_rs^r(b_rs^r+b_{r-1}s^{r-1}+\ldots+b_1s+1)+\ldots\\
&+m_{2r}s^{2r}(b_rs^r+b_{r-1}s^{r-1}+\ldots+b_1s+1)\\
&=a_{r-1}s^{r-1}+\ldots+a_1s+a_0+O(s^{2r})(b_rs^r+b_{r-1}s^{r-1}+\ldots+b_1s+1)
\end{aligned} \tag{3.22}$$

By matching the coefficients of s^i terms of (3.22) for $i=r,\ldots,2r-1$, we get the following equation system for the coefficients b_i:

$$\begin{bmatrix} m_0 & m_1 \ldots & m_{r-1} \\ m_1 & m_2 \ldots & m_r \\ \vdots & \vdots & \vdots \\ m_{r-1} & m_r & m_{2r-2} \end{bmatrix} \begin{bmatrix} b_r \\ b_{r-1} \\ \\ b_1 \end{bmatrix} = -\begin{bmatrix} m_r \\ m_{r+1} \\ \vdots \\ m_{2r-1} \end{bmatrix} \tag{3.23}$$

The coefficients a_i can be determined by matching the coefficients of s^i terms of Equation 3.22 for $i=0,\ldots,r-1$:

$$\begin{aligned}
a_0&=m_0\\
a_1&=m_0b_1+m_1\\
&\vdots\\
a_{r-1}&=m_0b_{r-1}+m_1b_{r-2}+\ldots+m_{r-2}b_1+m_{r-1}
\end{aligned} \tag{3.24}$$

However, computing Padé approximants using explicit moment computations, as is done in the asymptotic waveform evaluation algorithm [100], is numerically unstable [97]. It turns out that the numerically stable method for moments evaluation the so-called Krylov subspaces, defined as follows. For $r=1,2,\ldots$, the subspace

$$K_r^R\{A,b\}=span\{b,Ab,\ldots,A^{r-1}b\} \tag{3.25}$$

of R^n is called the rth *right Krylov subspace*, induced by A and b, and the subspace

$$K_r^L\{A^T,e\}=span\left\{e,A^Te,\ldots,(A^T)^{r-1}e\right\} \tag{3.26}$$

of R^n is called the rth *left Krylov subspace*, induced by A and b.

Note that moments can be computed from Equation 3.25 and Equation 3.26 by computing the inner products:

$$m_{2i} = -((A^T)^i e)^T \cdot (A^i b) \quad \text{and} \quad m_{2i+1} = -((A^T)^i e)^T \cdot (A^{i+1} b) \tag{3.27}$$

for $i = 0, 1, \ldots, r-1$. Unfortunately, the vectors from Equation 3.25 and Equation 3.26 quickly become almost linearly dependent [97] and there is a rapid accumulation of rounding errors. The solution is to construct more stable basis vectors v_1, $v_2, \ldots, v_r$ and $w_1, w_2, \ldots, w_r$ saved in columns of matrices W_r and V_r, such that $K_r^R\{A, b\} = span\{v_1, \ldots, v_r\}$ and $K_r^L\{A^T, e\} = span\{w_1, \ldots, w_r\}$.

There are two main approaches for constructing basis matrices W_r and V_r for Krylov subspaces, the Lanczos algorithm and the Arnoldi process.

3.2.1 Lanczos Algorithm

Given a matrix $A \in R^{n \times n}$, a right starting vector $b \in R^n$ and a left starting vector $e \in R^n$, the classical Lanczos reduction algorithm (process) generates a pair of biorthogonal basis $W_r, V_r \in R^{n \times r}$ for subspaces from Equation 3.25 and Equation 3.26, where

$$w_i^T v_j = 0 \text{ , for all } i \neq j \tag{3.28}$$

$v_1 = b/\|b\|$, $w_1 = e/\|e\|$ and $\|.\|$ is the 2-norm. It generates further a tridiagonal matrix $T_r \in R^{r \times r}$, which is related to the original system matrix A by

$$W_r^T A V_r = T_r \tag{3.29}$$

and can be considered an oblique projection of A onto the subspace from Equation 3.25 while remaining perpendicular to the subspace from Equation 3.26. Figure 3.2 illustrates the orthogonal and oblique projection of a vector.

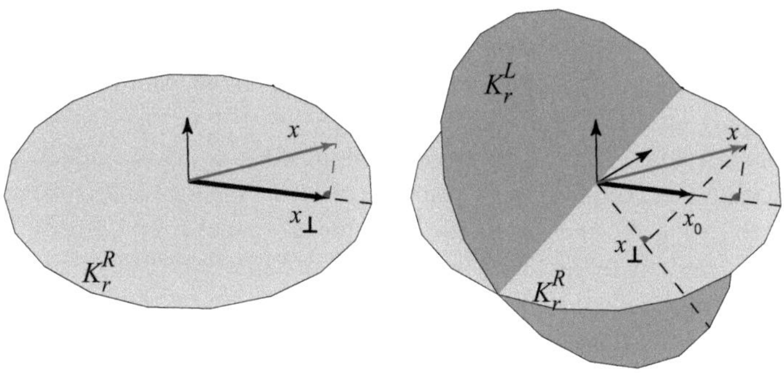

Fig. 3.2. Example of an orthogonal and an oblique projection of a vector, adapted from [6].

Note that T_r equals the system matrix of the reduced system A_r. An algorithm for the basic Lanczos process can be found in [97] or [101]. If the Lanczos algorithm is carried to the end with n being the last step, it can be viewed as a means of tridiagonalizing A by a similarity transformation:

$$V_n^{-1} A V_n = T_n \tag{3.30}$$

This allows one to write a transfer function, Equation 3.16, as:

$$\begin{aligned} G(s) &= -e^T (I - sV_n T_n V_n^{-1})^{-1} b \\ &= -e^T V_n (I - sT_n)^{-1} W_n^T b \\ &= -e^T b i_1^T \cdot (I - sT_n)^{-1} \cdot i_1 \end{aligned} \tag{3.31}$$

where i_1 is a first standard unit vector in R^n. Note that Equation 3.31 is due to Equation 3.28 and the fact that $v_1 = b/\|b\|$ and $w_1 = e/\|e\|$. The moments of $G(s)$ are given by

$$m_i = -e^T A^i b = -e^T b i_1^T \cdot T_n^i \cdot i_1 \tag{3.32}$$

The reduced-order transfer function $G_r(s)$ is defined by

$$\begin{aligned} G_r(s) &= -e_r^T (I - sA_r)^{-1} b_r \\ &= -e^T V_r (I - sT_r)^{-1} W_r^T b \\ &= -e^T b i_1^T \cdot (I - sT_r)^{-1} \cdot i_1 \end{aligned} \tag{3.33}$$

where T_r is the $r \times r$ leading principal submatrix of T_n and i_1 is a first standard unit vector in R^r. The moments of $G_r(s)$ are given by

$$m_{ri} = -e_r^T A_r^i b_r = -e^T b i_1^T \cdot T_r^i \cdot i_1 \tag{3.34}$$

Since $i_1^T \cdot T_n^i \cdot i_1 = i_1^T \cdot T_r^i \cdot i_1$ for $0 \le i \le 2r-1$ (this can be proved by mathematical induction [101]), we immediately see that the first 2r moments of $G(s)$ and $G_r(s)$ match, i.e. $G_r(s)$ is a Padé approximant of $G(s)$.

The main advantage of the Lanczos algorithm is that $G_r(s)$ is uniquely specified due to matching a maximal number of moments, $2r$. This means that we can apply the projection (using V_r and W_r as the basis for subspaces from Equation 3.25 and Equation 3.26) directly to the linear system with two matrices, Equation 2.14, or the right-hand side representation, Equation 3.1, and get the same reduced system. Furthermore, as the basic Lanczos algorithm is based on simple three-term recurrences (see [97] or [101]), it is reasonably fast for large r. However, it should be noted that in practical implementations, the basic Lanczos algorithm can break down for

$w_i^T v_i \approx 0$. This problem could be solved by using a much more complicated version of the Lanczos process, as suggested in [102]. Another disadvantage of the Padé approximants is that the reduced models generated may not be passive and stable in general, even though the original linear dynamic system was passive and stable [103]. Suggestions for improving the preservation of stability and passivity can be found in [104].

3.2.2 Arnoldi Process

The Arnoldi process produces only one sequence of vectors $v_1, v_2, \ldots, v_r$, which span the right Krylov subspace, Equation 3.25, and are othonormal:

$$v_i^T v_j = \begin{cases} 1, \text{if } i=j \\ 0, \text{if } i \neq j \end{cases} \text{ for all } 1 \leq i,j \leq r \tag{3.35}$$

It further generates an upper Hessenberg matrix $H_r \in R^{r \times r}$ (with $h_{ij} = 0, \forall (i,j), |i-j| > 1$), which is related to the system matrix A as follows:

$$V_r^T \cdot A \cdot V_r = H_r \tag{3.36}$$

The matrix H_r can be considered an orthogonal projection of the matrix A onto the Krylov subspace, Equation 3.25 (see Figure 3.2 left), and it is equal to the system matrix of the reduced system A_r. Algorithm 3.2 (see [97]) presents a complete statement of the Arnoldi process for the SISO setup.

Algorithm 3.2 *Arnoldi process*

Inputs: *Matrix* $A \in R^{n \times n}$ *and the starting vector* $b \in R^n$ *and order* r *of the reduced system*

Outputs: *Projection matrix* $V_r \in R^{n \times r}$ *and the (upper Hessenberg) system matrix* of the reduced system $A_r = V_r^T A V_r$

1. Set $\hat{v}_1 = b$.

For $i = 1, \ldots, r$ *do (build jth Arnoldi vector* v_j*):*

2. Compute $h_{i,i-1} = \|\hat{v}_i\|$.

If $h_{i,i-1} = 0$, *then stop.*

(The Krylov subspace $K_r^R\{A, b\}$ *is exhausted).*

3. Set $v_i = \hat{v}_i / h_{i,i-1}$.

4. Set $v_{i+1} = A v_i$.

5. For $j = 1, \ldots, i$ *do:*

Set $h_{j,i} = v_j^T \hat{v}_{i+1}$ *and* $\hat{v}_{i+1} = \hat{v}_{i+1} - v_j \cdot h_{j,i}$.

End_For.

End_For.

$V_r = [v_1 \dots v_r]$, $A_r = H_r = (h_{i,j}), i,j = 1, \dots r$.

The system Equation 3.15 is now reduced by projection:

$$A_r = V_r^T A V_r \ , \quad b_r = V_r^T b \ , \quad e_r = V_r^T e \tag{3.37}$$

In order to demonstrate the moment matching property of the Arnoldi process, we first emphasize the following properties:

$$A V_r = V_r H_r + h_{r+1,r} v_{r+1} i_r^T \tag{3.38}$$

$$A^k b = \|b\| V_r H_r^k i_1 \quad \text{, for } k = 0, \dots, r-1 \tag{3.39}$$

where i_1 and i_r are the first and the rth standard unit vectors in R^r. Equation 3.38 is obvious from Algorithm 3.2.

For property Equation 3.39 we note that since the entries of H_r below its lower second diagonal are all zeros, it can be proven by mathematical induction (we will not supply the evidence here) that the entries of H_r^k below its lower $(k+1)$ th diagonal are also all zeros. Hence,

$$i_r^T \cdot H_r^k \cdot i_1 = 0 \quad \text{for } k = 0, \dots, r-1 \tag{3.40}$$

From Equation 3.38 and the fact that $v_1 = b/\|b\|$, which implies $b = \|b\| V_r i_1$, we further obtain:

$$Ab = A \cdot \|b\| V_r i_1 = \|b\| V_r H_r i_1 + \|b\| h_{r+1,r} v_{r+1} i_r^T i_1 = \|b\| V_r H_r i_1 \tag{3.41}$$

From Equation 3.38 and Equation 3.40 (again by mathematical induction), it follows Equation 3.39. Now, the moments of $G(s)$ are defined as

$$\begin{aligned} m_k &= -e^T A^k b = -e^T \|b\| V_r H_r^k i_1 = \\ &= \underbrace{-e^T V_r}_{-e_r^T} \underbrace{H_r^k}_{A_r^k} \underbrace{\|b\| i_1}_{b_r} = m_{rk} \qquad \text{for } k = 0, \dots, r-1 \end{aligned} \tag{3.42}$$

Hence, in this case only r moments are matched and

$$G_r(s) = -e_r^T (I - sA_r)^{-1} b_r = -e^T V_r (I - sH_r)^{-1} \|b\| i_1 \tag{3.43}$$

is called a Padé-type approximant of $G(s)$. Consequently, the Arnoldi process is not invariant to system representation, i.e. if we apply the projection (using V_r as the basis for subspace from Equation 3.25) to Equation 3.1 or Equation 2.14 instead to Equation 3.15, we will get slightly different reduced order systems. Another disadvantage of the Arnoldi method is that each new Arnoldi vector should be orthogonal to all previously generated vectors. This means that the computational effort grows disproportionately to the dimension of the subspace. However, it can be shown (see [103]) that if we apply the Arnoldi-based projection to the first-order stable and passive ODE system, Equation 2.14, with symmetric and positive definite system matrices C and K, the reduced order model will also be stable and passive. So, in simple terms, the Arnoldi process trades some optimality (in the sense of matching as many moments as possible) to gain guaranteed stability and passivity.

3.2.3 Arnoldi versus Lanczos

In Table 3.2 the most important properties of both reduction algorithms are summarized.

Accuracy of approximation: The Lanczos algorithm produces a reduced system closer to the original one, because the number of moments matched here is twice that of the Arnoldi process. This has a simple explanation. Model reduction by the Arnoldi process does not take into account output vector e at all, while model reduction by means of the Lanczos algorithm is made by an oblique projection on the right Krylov subspace, Equation 3.25, that takes into account the left Krylov subspace, Equation 3.26. It is possible to increase the number of matched moments by Arnoldi to $2r$ by modifying it to so-called 2-sided-Arnoldi [105, 106], which takes into account the output vector. However, the algorithm becomes more complex and loses its ability to preserve the stability and passivity of the original system.

Computational complexity: One disadvantage of the Arnoldi method is that each new Arnoldi vector should be orthogonal to all previously generated vectors. This means that the computational effort for orthogonalization over the r steps of the algorithm grow as $O(2r^2n)$ with the dimensions n of the full space and r of the Krylov the subspace. Additionally, r steps of the Arnoldi procedure require r matrix–vector products at the cost of $2rNz(A)$, where $Nz(A)$ is the number of non-

Table 3.2. Comparison of Arnoldi and Lanczos

Properties	Arnoldi	Lanczos
Accuracy of approximation	r moments match	$2r$ moments match
Computational complexity	$O(2r^2n+2rN_z(A))$	$O(16rn+4rN_z(A))$
Invariance properties	no	yes
Numerical stability	yes	no
Preservation of stability and passivity	yes	no
Complete output approximation	yes	no

zero elements of A ($Nz(A)=n^2$ for a dense matrix). Thus, on average the computational effort for the Arnoldi algorithm grows as $O(2r^2n + 2rNz(A))$. The Lanczos algorithm requires less computational effort for orthogonalization. At each step it is necessary to deal with just two previously generated vectors (matrix T_r is tridiagonal), which makes the orthogonalization costs over the r steps of the algorithm grow only as $O(16rn)$. Together with matrix–vector products, the computational effort of the Lanczos algorithm are $O(16rn+4rNz(A))$. Hence, for large r it is faster than Arnoldi.

Invariance properties: Changing the representation of the original system does not change the input–output behavior of the reduced-order models generated by the Lanczos algorithm. The Arnoldi process is not invariant to system representation due to matching only r moments, which is half the number of unknowns in $G_r(s)$.

Numerical stability: Because of rounding errors in both algorithms, the column vectors of basis V and W for the Krylov subspaces from Equation 3.25 and Equation 3.26 may become non-orthogonal. How quickly this happens depends on the chosen dimension r of the Krylov subspaces. Since in the Arnoldi process each new vector should be orthogonal to all previously generated vectors, the rounding errors accumulate slower than by the Lanczos algorithm, where each new vector is orthogonalized only with respect to the last two generated vectors.

Preservation of the stability and passivity of the original system: The original dynamic system can be stable, that is, when time goes to infinity the values of x remain finite (bounded) and passive. Such a system does not generate energy. If so, then it is important that the reduced system also possesses these properties. Unfortunately, neither the Arnoldi nor the Lanczos algorithms in their basic form guarantee this and extra effort is needed to preserve the properties of the original dynamic system. As the Arnoldi process is mathematically simpler than the Lanczos algorithm (this is emphasized by their names: process and algorithm), in engineering applications it is more frequently used. The coordinate transformed Arnoldi [107] guarantees stable model reduction and the "block Arnoldi plus congruent transform" (PRIMA) from [108] guarantees passivity by transforming the system matrices so that they are positive semi-definite.

Approximation of the complete output: In general the Lanczos algorithm produces reduced-order models, which are "optimized" for particular output(s). This is because it takes into account the output vector c by using a basis W of the output Krylov subspace, Equation 3.26. The Arnoldi process, on the other hand, does not take into account the output vector at all, which enables the approximation of all outputs. In other words, it works even if the output matrix is a unity matrix, i.e. $E = I_{n\times n}$ in Equation 3.1 or Equation 3.15. From a control theory viewpoint, we refer to such a system as a single-input–complete-output (SICO). This aspect will be demonstrated numerically in Chapter 5.

Nevertheless, both approaches are based on moment matching and by nature are local because they accurately approximate the transfer function, Equation 3.16, only near the expansion point s_0. This can be improved by multi-point expansion, i.e. by expanding the transfer function about several points s_i and requiring the reduced transfer function to match the first moments at all expansion points. This idea has been implemented in the so-called rational Krylov methods [109–111]. The main methodological challenges in multi-point expansion methods are how to choose the expansion points and to determine how many of them are needed.

Lastly, let us note that in both approaches, instead of just one starting vector v, one can take a number of starting vectors expressed by the matrix B. This leads to a generalization of the Arnoldi and Lanczos algorithms to the so-called block-Arnoldi and block-Lanczos algorithms [97, 103], which are suitable for the reduction of multiple-input–multiple-output (MIMO) systems. Block Arnoldi will be introduced in Chapter 6.

3.3 Guyan Reduction

In the last section of this chapter we describe a commercially available [112] Guyan reduction, which has been commonly used in mechanical engineering for many years and can easily be applied to the thermal domain as well. The Guyan algorithm [11] is another MOR method which projects (similar to Equation 2.36 and Equation 2.37) a high-dimensional ODE system to a lower-dimensional one. A projection subspace is chosen, however, based on engineering intuition rather than on mathematical rigor. Practically, this means that the designer can choose the "important" finite element nodes, which are to be physically preserved within a reduced model.

3.3.1 Static Matrix Condensation

The system matrices obtained by spatial discretization of the heat transfer PDE contain terminal nodes which connect to the external circuitry as well as to internal nodes. The large dimension of Equation 2.14 could be effectively reduced by the elimination of internal nodes. For steady-state problems of the form:

$$K \cdot T = F \cdot u(t) \tag{3.44}$$

it is possible to decompose the linear system in Equation 3.44 into terminal and internal equations by splitting the matrix K into four blocks:

$$\begin{bmatrix} K_{ee} & K_{ei} \\ K_{ie} & K_{ii} \end{bmatrix} \cdot \begin{bmatrix} T_e \\ T_i \end{bmatrix} = \begin{bmatrix} F_e \\ F_i \end{bmatrix} \tag{3.45}$$

with the index sets e and i ranging over all external and internal nodes respectively. For simplicity, the input term $u(t)$ has been incorporated in the load vector. It is further possible to eliminate the equations for the non-terminal nodes by means of linear algebra operations (e.g. the Schur complement) [113] as follows.

By partitioning the temperature and load vectors consistently with Equation 3.45, the system in Equation 3.44 can be decomposed as

$$F_e = K_{ee} \cdot T_e + K_{ei} \cdot T_i \tag{3.46}$$

$$F_i = K_{ie} \cdot T_e + K_{ii} \cdot T_i \tag{3.47}$$

In order to eliminate the equations for the internal nodes, one can express T_i from Equation 3.47 as

$$T_i = K_{ii}^{-1} \cdot (F_i - K_{ie} \cdot T_e) \tag{3.48}$$

and insert it into Equation 3.46 to obtain

$$(K_{ee} - K_{ei} \cdot K_{ii}^{-1} \cdot K_{ie}) \cdot T_e = F_e - K_{ei} \cdot K_{ii}^{-1} \cdot F_i \tag{3.49}$$

The dimension of the reduced system in Equation 3.49 is determined by the number of terminal nodes chosen. It is not necessary to compute K_{ii}^{-1} explicitly, but rather the LU decomposition of K_{ii} and back substitutions corresponding to each matrix–vector product can be made. Additional speed-up is due to the sparse structure of K_{ii}. The reduced heat conductivity matrix and load vector are then defined as

$$\begin{aligned} K_r &= K_{ee} - K_{ei} \cdot K_{ii}^{-1} \cdot K_{ie} \\ F_r &= F_e - K_{ei} \cdot K_{ii}^{-1} \cdot F_i \end{aligned} \tag{3.50}$$

Please note that until now no approximation has been used and hence the condensation of the heat conductivity matrix is exact.

3.3.2 Dynamic Matrix Condensation

Generalizing the elimination of internal nodes via the Schur complement to the dynamic problems was first proposed by Guyan [114]. It was originally applied only to undamped structural analysis. Starting from the static structural equation, Equation 3.45, Guyan assumed that no loads were applied on the internal nodes ($F_i = 0$). In this case, after replacing the temperature vector with the displacement vector, the Equation 3.48 changes to

$$x_i = -K_{ii}^{-1} \cdot K_{ie} \cdot x_e \tag{3.51}$$

and the Schur complement amounts to a coordinate transformation of the form $x = V_G \cdot x_e$ with

$$V_G = \begin{bmatrix} I \\ -K_{ii}^{-1} K_{ie} \end{bmatrix} \tag{3.52}$$

Here, I is the unity matrix, whose dimension corresponds to the number of terminal nodes. One can state based on Equation 3.50 that the reduced heat conductivity (stiffness) matrix and the load vector are built by projection:

$$\begin{aligned} K_r &= V_G^T \cdot K \cdot V_G \\ F_r &= V_G^T \cdot F \end{aligned} \tag{3.53}$$

Guyan further employed the transformation from Equation 3.52 to the kinetic and potential energies of the structure, defined as

$$E_k = \frac{1}{2}\dot{x}^T \cdot M \cdot \dot{x}$$
$$E_p = \frac{1}{2}x^T \cdot K \cdot x \tag{3.54}$$

to get:

$$E_k = \frac{1}{2}\dot{x}_e^{\,T} \cdot V_G^T \cdot M \cdot V_G \cdot \dot{x}_e$$
$$E_p = \frac{1}{2}x_e^{\,T} \cdot V_G^T \cdot K \cdot V_G \cdot x_e \tag{3.55}$$

Again, one can state, based on Equation 3.55, that the reduced mass matrix is built by projection:

$$M_r = V_G^{\,T} \cdot M \cdot V_G \tag{3.56}$$

or in terms of partitioning it consistently with Equation 3.45:

$$M_r = M_{ee} - M_{ei}K_{ii}^{-1}K_{ie} - (K_{ii}^{-1}K_{ie})^T(M_{ie} - M_{ii}K_{ii}^{-1}K_{ie}) \tag{3.57}$$

There is no mathematical proof for approximation properties of Equation 3.57 but rather the analogy to Equation 3.53 is used to derive it.

There has been a number of attempts to modify the Guyan algorithm for the damped structural problems, e.g. [115] and [116]. The commercial FE solver ANSYS [112] also offers the possibility of Guyan-based reduced order modeling for transient heat transfer problems, Equation 2.14. The computation of K_r and F_r is done as in Equation 3.50 and the reduced heat capacity matrix is given through

$$C_r = C_{ee} - C_{ei}K_{ii}^{-1}K_{ie} - K_{ei}K_{ii}^{-1}C_{ie} + K_{ei}K_{ii}^{-1}C_{ii}K_{ii}^{-1}K_{ie} \tag{3.58}$$

which is again similar to Equation 3.57, assuming a symmetric K matrix and therefore $K_{ie}^T = K_{ei}$ and $(K_{ii}^{-1})^T = K_{ii}^{-1}$. Now a reduced thermal system has a form:

$$C_r \cdot \dot{T}_e + K_r \cdot T_e = F_r \tag{3.59}$$

with $K_r = V_G^{\,T} \cdot K \cdot V_G$, $C_r = V_G^{\,T} \cdot C \cdot V_G$ and $F_r = V_G^{\,T} \cdot F$. Even for the case where $F_i \neq 0$ the projection matrix V_G is used as defined in Equation 3.52. After the reduction, it is possible to expand the terminal degree of freedom values to gain the complete temperature distribution of the device using Equation 3.48 for the steady-state.

The main advantage of Guyan-based MOR is that it physically preserves the terminal nodes, as schematically shown in Figure 3.3. This is advantageous for the coupling of several reduced order models as will be demonstrated in Chapter 6.

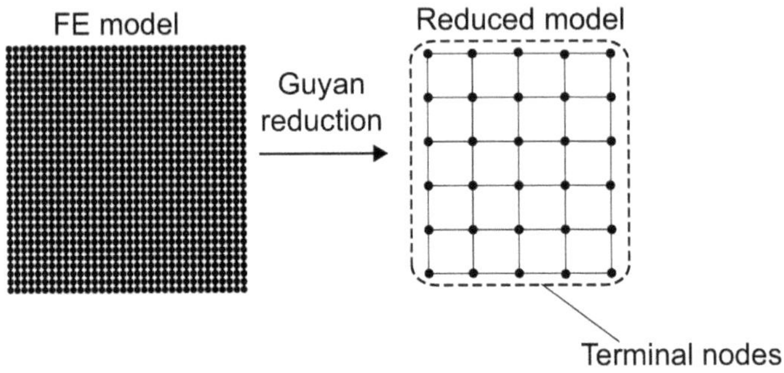

Fig. 3.3. Schematics of Guyan-based MOR with physical preservation of the chosen terminal nodes.

However, the modified Guyan method will provide sufficient accuracy for the reduction of thermal systems only if the number of master degrees of freedom chosen is large enough (see numerical results in Chapter 5). There are no strict guidelines on how to choose the master degrees of freedom.

3.4 Summary

In this chapter we have presented the most important methods for MOR of linear dynamic systems. These methods and their most important properties are summarized in Table 3.3.

Table 3.3. Methods for MOR of linear dynamic systems

Method	Advantages	Disadvantages
Control theory methods (balanced truncation approximation, singular perturbation approximation, Hankel norm approximation)	have a global error estimate, can be used in a fully automatic manner	computational complexity is $O(n^3)$, hence can be used only for systems with less than a few thousand unknowns
SVD-Krylov (low-rank Grammian approximants) and matrix sign function methods	have a global error estimate and the computational complexity is less than $O(n^2)$	currently under development
Padé approximants (moment matching) via Krylov subspaces by means of either the Arnoldi or Lanczos algorithm	computationally very advantageous, can be applied to very high-dimensional first-order linear systems	do not have a global error estimate. Hence, it is necessary to select the order of the reduced system manually
Guyan-based methods	preserve the physical nodes	result in unnecessary large reduced order models

Control theory methods are mathematically optimal, i.e. offer a global error estimate of the difference between the transfer function of the original high-dimensional and reduced low-dimensional system. They show that a model reduction problem of the linear dynamic system is solved in principle, but requires a computational complexity of $O(n^3)$, and therefore is only applicable to small linear systems. The SVD-Krylov methods (SVD: singular value decomposition), which are based on low-rank Grammian approximants [117–120], or the matrix sign function methods [121] have resulted from efforts to find computationally effective strategies in order to apply control theory methods to large-scale systems. Up to now, most of the practical work in model reduction of large linear dynamic systems has been tied to Padé approximants of the transfer function via Krylov subspaces by means of either the Arnoldi or the Lanczos process. These methods have no global error estimate, but can easily be applied to high-dimensional ODE systems. Commercially available algorithms based on modification of Guyan's method preserve the physical nodes but result in unnecessarily large reduced order models.

Due to its mathematical simplicity and numerical stability we think that at present the Arnoldi process is the most suitable tool for MOR of electro-thermal engineering problems. The numerical results in Chapter 5 will prove this.

4

Selected Model Reduction Software

In the previous two chapters we have described different possibilities for the solution of the heat transfer PDE and we have given an overview of the most important methods for linear MOR. The goal of this chapter is to stress that model reduction software is available and it is already possible to use model reduction in everyday practical work.

The choice of software discussed in this chapter is, no doubts, subjective and reflects our own computational environment. Nonetheless, we hope that this will help you to start using model reduction in your own work. The software is described in a step-by-step tutorial manner and we use a microthruster to demonstrate its functioning. All files mentioned in this chapter are available from the `MOR for ANSYS` site (`http://www.imtek.uni-freiburg.de/simulation/mor4ansys/`). Additionally you will need ANSYS to generate a high-dimensional finite element model and Mathematica to compare simulation results of the original and reduced systems. MOR for ANSYS functions with the traditional ANSYS Inc. solver products such as ANSYS Structural, ANSYS Mechanical, ANSYS Multiphysics and the ANSYS University versions of ANSYS Multiphysics. Note that MOR for ANSYS does not function with the ANSYS Workbench environment, ANSYS CFX or ANSYS ICEM CFD products. If you do not have Mathematica, you can work with a reduced model in Matlab (a script to read a reduced model to Simulink is available from the MOR for ANSYS site) or any other environment but you may need to write your functions. If you have finite element software other than ANSYS, you can still use MOR for ANSYS but it is necessary to learn how to obtain the system matrices from the software.

In Section 4.1, the SLICOT library, which includes gramian-based model reduction methods, is introduced. The software MOR for ANSYS, which implements implicit moment matching via the Arnoldi process, is presented in Section 4.2. In Section 4.3, we present our Mathematica functions (`Post4MOR`) to work with a reduced model and a Mathlink interface to the SLICOT library.

4.1 SLICOT Library

Model reduction methods developed in the framework of control theory (see Section 3.1) are in a matured state and there are a number of tools implementing them, for example, MATLAB Control System Toolbox, MATCONTROL, MATRIX_X, SLICOT etc. A good description of many such packages can be found in [9].

In our own practice we have used SLICOT [123, 13], which stands for Subroutine Library in Control Theory. For academic research, it is freely available at http://www.slicot.de. This is a collection of Fortran 77 subroutines using in turn subroutines from BLAS (Basic Linear Algebra Subroutines) and LAPACK (Linear Algebra Package) [124] for numerical linear algebra. There exists a parallel version of SLICOT that uses ScaLAPACK [125] for parallel linear algebra. Fortran subroutines AB09AD, AB09BD and AB09CD implement balanced truncation approximation, singular perturbation approximation and Hankel-norm approximation respectively. The library also contains other mathematical tools such as discrete sine/cosine and Fourier transformations. A full list of SLICOT subroutines, the documentation and examples are available on the SLICOT web page.

One can call SLICOT subroutines directly from Fortran, C or C++. Alternatively, there is a MATLAB interface to SLICOT that allows a user to call subroutines directly from the MATLAB environment. We have implemented a Mathlink interface (follow a link from the SLICOT site) that allows a user to call model reduction subroutines directly from the Mathematica environment. It will be described in more detail in Section 4.3. The binary version of our Mathlink interface for Windows is available from the SLICOT site. The numerical results for model reduction via control theory methods presented in Chapter 5 have also been obtained by using SLICOT.

We have chosen SLICOT not only because of its free availability but also due to its reliability and high performance as well as its parallel extension. A comparison of SLICOT and other control theory model reduction tools is given elsewhere (see Tables 7.2 and 7.3 in [13]). However, as was mentioned in Section 3.1, the computational complexity to solve the Lyapunov equations and find Hankel singular values is of $O(N^3)$. Table 4.1 displays the computational time for model reduction of a simple thermal MEMS as a function of the problem dimension. One can see that a two-fold increase in dimension increases the computation time by about eight times for both serial and parallel versions. Hence in spite of its unique performance, the library can be efficiently used for models of moderate size only. Additional problem may come from the SLICOT memory requirements. As with most other control theory tools, it expects to have system matrices in the dense form and does not allow one to exploit the sparse structure of matrices arising from finite element models.

4.2 *MOR for ANSYS* – Model Order Reduction for ANSYS

Implicit moment matching described in Section 3.2 allows us to preserve the sparse structure of system matrices arising from the finite element method and, as a result, to

Table 4.1. Computational time in seconds for model reduction with SLICOT on 450 MHz Sun Ultra-80 with 4Gb RAM and four processors

Dimension	Time serial	Time, parallel (4 processors, 2x2)
600	60	25
1332	703	130
2450	4346	666
3906		2668

treat high-dimensional models. However, when we started research on model reduction in 2001, one could find the implementation of the Arnoldi process in MATLAB only. Then we decided to make a scalable implementation in C++ and the result of our efforts was MOR for ANSYS [12], short for model order reduction for ANSYS. In our experience, on a computer with 4 Gb of RAM, MOR for ANSYS allows us to perform model reduction for finite element models with up to 500 000 degrees of freedom.

The MOR for ANSYS structure is shown schematically in Figure 4.1. It is a stand-alone command-line program. It can read the system matrices directly from EMAT and FULL binary files made by ANSYS after the discretization of a computational domain. Alternatively, it can read in an arbitrary linear ODE system provided the system matrices have been supplied in the Matrix Market format [126] (`http://math.nist.gov/MatrixMarket/`). These are plain ASCII files with quite simple formatting rules. At the Matrix Market site, one can find the code in Fortran, C and MATLAB to read and write matrices in the Matrix Market format.

After the system matrices have been converted to the internal format, MOR for ANSYS runs the Arnoldi process to find a low-dimensional subspace and then projects the original system onto it. MOR for ANSYS uses high-performance direct solvers TAUCS [127] and UMFPACK [128, 129] based on ATLAS [130].

The source code in C++ contains about 4000 lines (as of version 1.8) and together with compiled binaries for Windows it is freely available under http://www.imtek.uni-freiburg.de/simulation/mor4ansys/.

MOR for ANSYS writes a reduced model in the Matrix Market format that makes it independent from the environment to work with reduced models. We describe in the next section a package Post4MOR designed to work with in Mathematica (follow the link from the MOR for ANSYS site to obtain the code). Yet, as already mentioned, one can relatively easily read the reduced model into MATLAB and Simulink or other interactive environments.

We have prepared a small step-by-step tutorial to demonstrate how one can use MOR for ANSYS in practice and it is briefly presented below. The full tutorial with all the files is available from the MOR for ANSYS site. We consider a thermal model of MEMS microthruster, the engineering problem described in Chapter 5.1. The model has about 20,000 nodes but, provided that your computer is fast enough, you can switch an element type within an ANSYS script to obtain a model with about 80,000 nodes. After you complete the tutorial, you will be able to perform model reduction with MOR for ANSYS for your own finite element models.

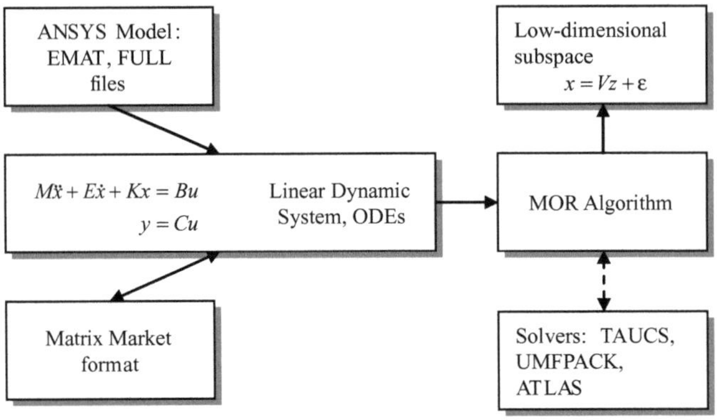

Fig. 4.1. MOR for ANSYS structure.

Step 1. Building Finite Element Model in ANSYS

Before model reduction, one has to make a finite element model. We provided an APDL (ANSYS Programming Development Language) script `model.ans` that will build a finite element model for the microthruster. This script should be executed and produce a `file.db`[1] file, which is the main ANSYS database.

ANSYS can be started from a command-line. This makes it more convenient to write the sequence of commands that you are supposed to make. In the tutorial, there is a csh script `runansys` to start ANSYS from a console window. However, one has to change the path to his/her own ANSYS bin directory and the name of the license server. `runansys` takes as an argument the filename to execute without an extension

```
./runansys model
```

and redirects ANSYS output to model.log. The command starts ANSYS as a background job from the current directory. If you find the ANSYS graphic user interface more convenient, change the current directory to that in which the file is located, then select File and Read Input From.

One text file prepared by a user is required in order to run `MOR for ANSYS`. It should specify output degrees of freedom. For a thermal problem, it should contain the node name (an arbitrary text without spaces) and the node number for each output. Alternatively, the option `-f` can be employed that will forces `MOR for ANSYS` to generate the complete output for all degrees of freedom. Numerical results on the approximation of the complete output after the Arnoldi-based MOR can be found in Chapter 5. However, to start with, we would recommend using a few outputs only. Our script `model.ans` creates file `output.txt` with four output nodes automatically.

[1] If the user does not specify another name, all the files created by ANSYS will be per default named file.*. This can be changed by /FILNAME command.

Step 2. Running Transient Simulation in ANSYS

It is unnecessary to simulate the full-scale model in order to run MOR for ANSYS, but it is useful to compare the results of the reduced order model with those of the original one when you learn model reduction. This will allow you to compare the time to integrate a finite element model in ANSYS with the time for model reduction and integration of a reduced model in Mathematica.

However, as numerical integration over time of the full-scale model in ANSYS leads to numerical errors, it is important to know the magnitude of numerical integration error before comparison with a reduced model. One way is to compare different integration strategies in ANSYS. In the tutorial, there are three options to integrate the full model over time: fixed timesteps, adaptive integration and timesteps varying in time logarithmically. The latter enjoys widespread use in electro-thermal simulation. There are three ANSYS scripts that should be executed: fixstep.ans, adaptive.ans and logtime.ans. They all need the database file.db from step 1 as well as the script simulation.ans (also available in the tutorial). The scripts can also be run in the batch mode as follows:

```
./runansys fixstep

./runansys adaptive

./runansys logtime
```

or from the graphic user interface. It should be noted that transient simulation in ANSYS takes time. Please be patient.

Our scripts will write ANSYS results to the text files `fixstep.txt`, `adaptive.txt`, and `logtime.txt` in the Matrix Market format. The result files contain the numerical values of the temperatures at the specified output nodes and we will use them in Section 4.3. Figure 4.2 shows the step response at two different output nodes when using fixed and logarithmic time stepping. The plots have been generated in Mathematica with package `Post4MOR` (see the next section).

Step 3. Model Reduction with MOR for ANSYS

As already mentioned, MOR for ANSYS needs binary ANSYS files `file.emat` and `file.full` with system matrices in order to assemble an ODE system as follows:

$$\begin{aligned} M\ddot{x}+E\dot{x}+Kx&=Bu \\ y&=C\cdot x \end{aligned} \tag{4.1}$$

for second-order models (structural mechanics) or

$$\begin{aligned} E\dot{x}&=Ax+Bu \\ y&=C\cdot x \end{aligned} \tag{4.2}$$

for first-order models (like thermal models). Here x is a vector of unknowns that includes all degrees of freedom, M, E and K are the system matrices, B is the input, C the output matrix, u is a vector of inputs and y is a vector of outputs. It should be

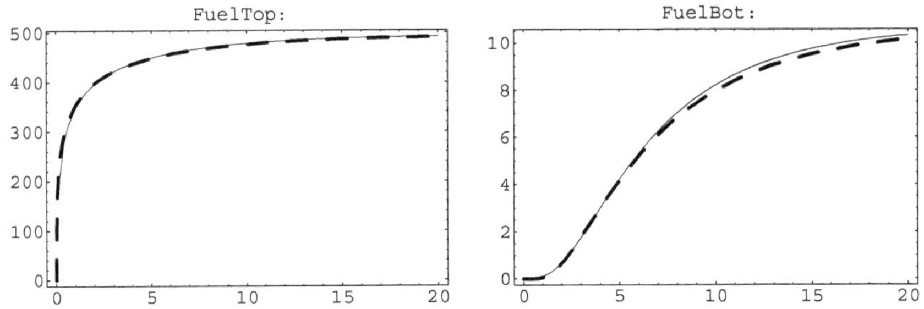

Fig. 4.2. Comparison of integration with fixed timesteps (full line) and with logarithmic timesteps (dashed line) in ANSYS. Step response (temperature versus time) at two different outputs.

noted that `MOR for ANSYS` naming convention complies with that for model reduction benchmarks [122] used by mathematicians. Matrix C is considered to be the output matrix and not the damping matrix as usual for the finite element convention. For first-order systems, the notation in Eq. 4.2 is also different from the notation in Eq. 2.14 $A = -K$. Please be careful.

Use the script matrices.ans to generate file.emat and file.full from the database file:

```
./runansys matrices
```

Note that the script must be executed immediately before using MOR for ANSYS. The `file.full` after the transient simulation cannot be used for model reduction as it contains a combination of system matrices required to solve a system of linear equations for a particular timestep.

The next command performs MOR with MOR for ANSYS

```
mor_for_ansys file.full file.emat -C output.txt
```

The dimension of the reduced model is by default 30, which can be changed with the option `-N`. A number of different options to control the model reduction process are described in the MOR for ANSYS manual.

MOR for ANSYS builds the projection matrix V_r by implicit moment matching via the Arnoldi process. For first-order models, the system is first shifted to the left-hand side formulation of Eq. 3.15 and then Algorithm 3.2 is used. For second-order models there are three options to construct matrix V_r (see the MOR for ANSYS manual). In each case the compact models are obtained by the projection of the original equations as follows:

$$\begin{array}{c} M_r\ddot{x} + E_r\dot{x} + K_r x = B_r u \\ y = C_r \cdot x \end{array} \quad \text{and} \quad \begin{array}{c} E_r\dot{x} = A_r x + B_r u \\ y = C_r \cdot x \end{array} \tag{4.3}$$

where $M_r = V_r^T M V_r$, $E_r = V_r^T E V_r$, $K_r = V_r^T K V_r$, $A_r = V_r^T A V_r$, $B_r = V_r^T B$, $C_r = C V_r$.

The system matrices for the reduced model is by default written to the files `mor.*` (for the microthruster model these are `mor.E`, `mor.A`, `mor.B` and `mor.C`) in the Matrix Market format. The base filename can be changed with the option `-o`.

4.3 Using Package Post4MOR and Mathlink Interface to SLICOT in *Mathematica*

We use Mathematica as an interactive computational environment. The package `Post4MOR` contains Mathematica functions to make transient and harmonic simulations with a reduced model and visualize results. The tutorial contains a Mathematica notebook plots.nb with step-by-step commands to load the reduced model, run transient simulation and compare results with ANSYS results. It requires to load functions from the file `Post4MOR.m` at the beginning and the files with ANSYS results `*.txt` and the reduced model `mor.*` obtained in the previous section. The commands within plots.nb should be executed successively. The notebook can also be considered as an introduction to various functions defined in `Post4MOR.m` like `ReadResult`, `ReadSystem`, `TransientSolution` and `PlotResult`. The complete documentation for `Post4MOR.m` is in an Appendix. In Figure 4.3 the step responses at two specified outputs of a thermal microactuator are shown.

As discussed previously, the SLICOT subroutines can be used for the reduction of moderate size models. Hence, it is possible to use them to further compact the model that has been obtained by MOR for ANSYS. We have developed a Mathematica interface to model reduction subroutines of the SLICOT library in order to seamlessly use them with Post4MOR functions. The library is interfaced via MathLink[2] and the notebook `SlicotManual.nb` gives guidelines how to use it from within Mathematica, and demonstrates how the reduction is working with the microthruster example. Figure 4.4 shows the hierarchy of the function calls and the libraries to be linked. The user can use the binary interface for Windows available from the SLICOT site or should compile SLICOT (together with BLAS and LAPACK).

The Mathlink interface should be installed in Mathematica before it can be used, as follows:

```
Install["path/slicot"]
```

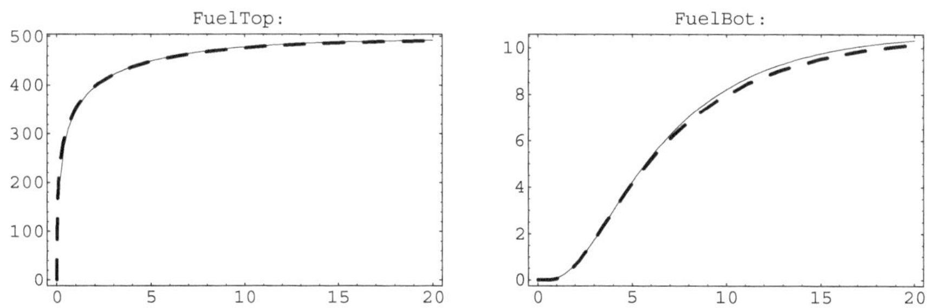

Fig. 4.3. Step response (temperature versus time) of the full-scale (20,360 DOF, full line) and the reduced-order 30 model (dashed line), at two different outputs.

[2]MathLink is an open interprocess communication protocol that allows external programs to call Mathematica or be called by Mathematica. It is a part of the Mathematica environment.

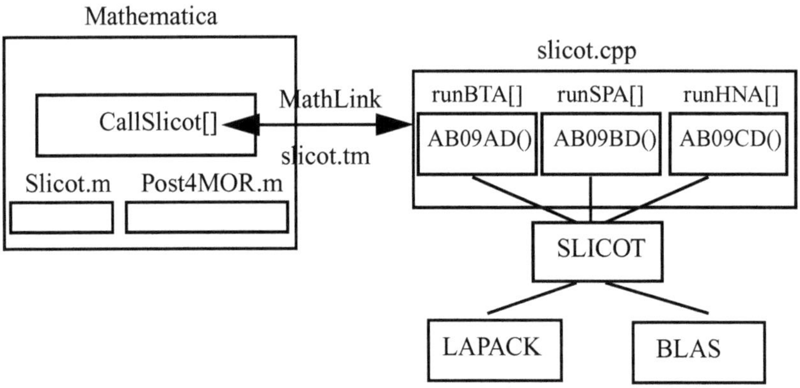

Fig. 4.4. Structure of the Mathematica interface for SLICOT.

(path should be replaced by the real location of the executable). After that, a new Mathematica function `CallSlicot` becomes available to a user. Formally, it looks like any other Mathematica function but when called it passes arguments via Mathlink to SLICOT subroutines. `CallSlicot` takes as arguments the first-order dynamic system in explicit form and either the desired dimension for the reduced system or the global error according to Eq. 3.6, Eq. 3.10 or Eq. 3.14. By default the truncated balanced approximation is found, but the user can switch to other two methods by setting an option *Method→SPA* or *Method→HNA*.

No doubt, the Mathematica interface for SLICOT can be used independently of MOR for ANSYS. One can enter any system as an argument to the function. `CallSlicot` returns the reduced system and the Hankel singular values of the original system. In Figure 4.5, the Hankel singular values of the reduced model of dimension 30 obtained by MOR for ANSYS from the original microthruster model are plotted. One can observe that even for the reduced model, the Hankel singular values decay rapidly. This shows the possibility of reducing its dimension even further.

The technique of using SLICOT to further compact reduced models obtained by MOR for ANSYS is discussed in Section 6.1 where we refer to it as consecutive model reduction. This is possible because a reduced model obtained by implicit moment matching is not optimal. On the other hand, it is impossible to employ SLICOT for the original high-dimensional model of a microthruster.

One can use `Post4MOR` functions to visualize and compare the results. Figure 4.6 shows the frequency responses of a thermal system of dimension 30 and two reduced systems computed by SLICOT (BTA, dimensions 5 and 10). Frequency responses were computed with the function HarmonicSolution.

It is worth noting that it is possible to run a Mathematica notebook and SLICOT on two different computers as Mathlink allows to connect different applications through network.

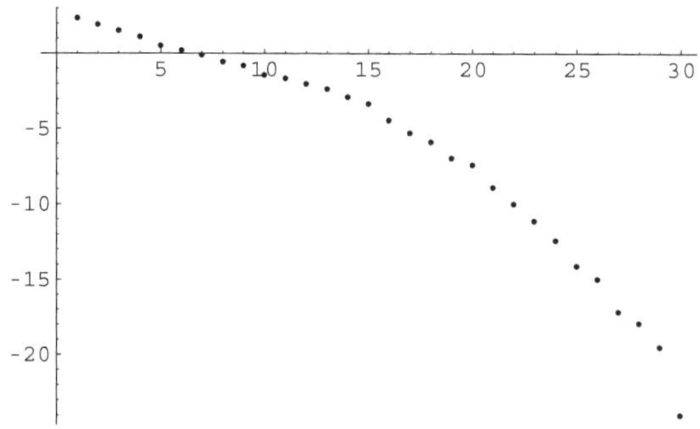

Fig. 4.5. Logarithm of Hankel singular values for the thermal model of dimension 30 obtained by MOR for ANSYS from the microthruster model.

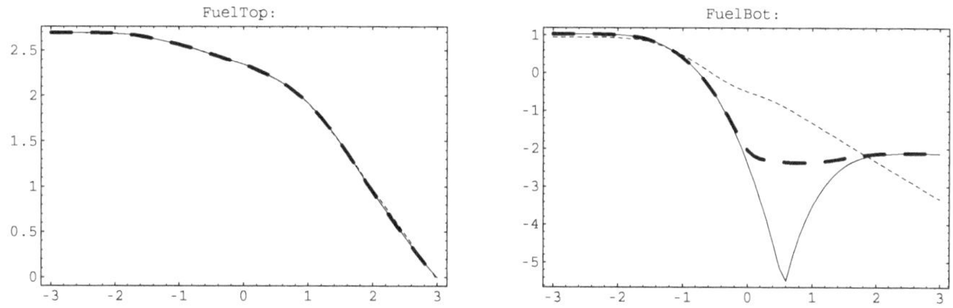

Fig. 4.6. Frequency response (transfer function versus logarithm of frequency) of a thermal model of order 30 (full line), reduced order 10 model (thick dashed line) and order 5 model (dotted line) at two different output nodes. Reduction was done with SLICOT (BTA).

4.4 Summary

The main goal of this chapter was to show that model reduction software is already available. It may take time to learn it and set it up for your particular environment but it is possible to incorporate results from Chapter 5 in real design right now.

The documentation for MOR for ANSYS, `Post4MOR` and Mathematica interface for SLICOT can be found in the Appendix.

5

Application of Model Reduction to Electro-thermal MEMS

In this chapter we describe the application of MOR to several electro-thermal MEMS models. It is important to understand, that although all the models presented belong to microhotplate-based devices with electrical excitation, the MOR is applied solely to the decoupled and discretized heat transfer equation (Eq. 2.14). Hence, the methodology presented can be applied to a spatially discretized thermal subsystem of any MEMS model, provided the decoupling (as described in Section 2.4) has been performed.

In Section 5.1 we present three novel MEMS devices used as case studies for MOR in our work. These are a pyrotechnical microthruster, a thermally tunable optical filter and a microhotplate gas sensor. Detailed device descriptions and the results of numerical simulations with ANSYS are presented. Section 5.2 demonstrates the decoupling of the electrical and thermal domains, which enables the application of the Arnoldi-based MOR. We would like to emphasize once more that at present we consider the Arnoldi process as the most suitable (both simple to implement and numerically stable) tool for MOR of electro-thermal MEMS. In Section 5.3 the different aspects of Arnoldi-based MOR, such as the approximation of the complete output, the reduction of systems with nonlinear input functions, the system-level simulation of the reduced model and the increase in computational efficiency, are numerically demonstrated. Section 5.4 compares the results of Arnoldi-based MOR with reduction using the Guyan algorithm (via substructuring in ANSYS) and reduction using control theory methods. Section 5.5 demonstrates the solution of the inverse thermal problem, which means the extraction of the material thermal parameters, via MOR and optimization. Section 5.5 summarizes the most important characteristics of the Arnoldi algorithm, when applied to electro-thermal MEMS models.

5.1 MEMS Case Studies

5.1.1 Pyrotechnical Microthruster

Pyrotechnics is a technology that is several centuries old and was mainly developed for military needs. The trend towards system miniaturization, however, has resulted in a completely new technology called micropyrotechnics. Promising applications of this technology are the medical field, security systems and especially space.

In recent years considerable interest in MEMS devices has arisen within the space industry [131]. Building a cluster of microsatellites (20–100 kg) or nanosatellites (< 20 kg) should be cheaper and more robust than building a single big satellite. These microspacecraft could be used in various space research fields, such as asteroid missions, disaster and magnetospheric monitoring, inspection of other spacecraft [132, 133] etc. A key point in the miniaturization of spacecraft is the micropropulsion subsystem, since in all the mentioned applications maneuvering plays a central role. Micropropulsion subsystems should be able to deliver very small (a few micronewtons to a few hundred millinewtons) and accurate forces needed for the stabilization, the pointing and station keeping of small satellites. In recent years, a number of micropropulsion systems using MEMS technology, such as cold gas microthrusters [134], subliming solids microthrusters [135], vaporizing liquid microthrusters [136], micropulsed plasma thrusters [137], microion thrusters, bi-propellant [139], and solid propellant microthrusters [140], have been investigated.

The short duration of the thrust impulse makes pyrotechnical solid propellant microthrusters most suitable for achieving the low velocity increments needed, for example, for station keeping of small satellites (short duration mission). In the last few years, a number of MEMS research groups from Europe, USA and Asia have been working on the development of these devices [141–147]. Their concept is based on the high rate combustion of one single propellant stored in a combustion chamber. The gas generated by combustion of the propellant is accelerated in a nozzle, thus delivering thrust. The concept of solid propellant microthrusters offers several advantages: there is no liquid fuel and hence no leakage, there are no moving parts and solid propellant ignition consumes little energy.

The solid propellant microthruster presented here was developed within a European project Micropyros (founded under IST-99047). It integrates solid fuel with four bonded silicon micromachined wafers [148] and delivers an impulse-bit thrust within a submillimeter volume of silicon by producing a large amount of energy from the ignitable substance contained within the microsystem. Microthruster fuel is ignited by passing an electric current through a polysilicon resistor embedded in a thin dielectric membrane, as shown in Figure 5.1. After the ignition phase, sustained combustion takes place and forms a high-pressure, high-temperature gas mixture. Under the pressure of the gas the membrane ruptures and an impulse is imparted to the carrier frame as the gas escapes from the tank. The lack of the restart ability is compensated for by the fabrication of an array structure. The dimensions

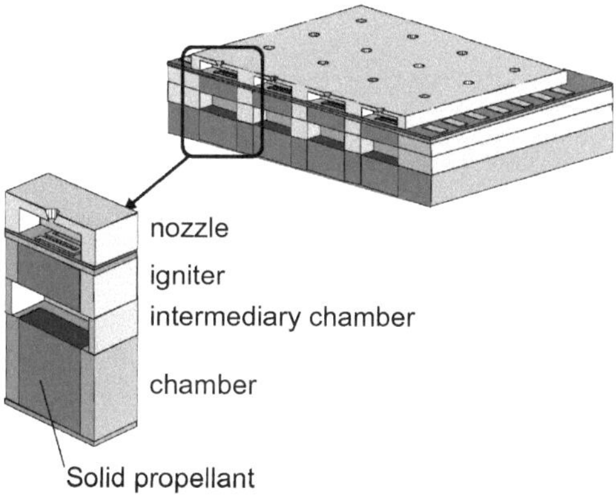

Fig. 5.1. Structure of a pyrotechnical microthruster array. Picture courtesy of C. Rossi (LAAS-CNRS, France).

of individual rockets are in the millimeter scale for the chamber and in the micrometer scale for the nozzle.

Igniters, nozzles and intermediary wafers are fabricated using silicon micromachining and combustion chambers are fabricated using Foturan glass or silicon. A fabrication of the igniter wafer is done by thermally oxidizing a silicon wafer and coating it with silicon rich LPCVD (low pressure chemical vapor deposition) nitride. In a third step, a polysilicon layer is deposited by LPCVD at 605°C and doped by diffusion. The heater filament is patterned using reactive ion etching (RIE). The electrical pads and electrical supply lines are fabricated in gold. To fabricate the membrane, the silicon is etched away by deep reactive ion etching (DRIE). The nozzles are also fabricated by DRIE. Two chamber materials are used: silicon and Foturan, which is a photostructurable glass that can be anisotropically wet etched. The main advantages of using silicon lie in its well-known fabrication technology and its high melting point. The silicon chambers were realized by DRIE. On the other hand, Foturan offers a lower heat conductivity that can result in better thermal insulation between the chambers, which could be of importance when considering the thermal cross-talk between a single rocket and its closest neighbors.

A complete modeling strategy for a microthruster array combines an electrical circuit, heat transfer, combustion, membrane rapture and gas dynamics. It happens, however, that the main modeling aspects must include electro-thermal simulation and a system-level simulation of the array and the driving circuitry. All other aspects of microthruster operation can be modeled at the basic level, as suggested in [149].

Here, we consider the initial heating phase of the fuel, right up to the onset of ignition. Electro-thermal simulation helps to design a device in such a way as to reach the ignition temperature within the fuel and at the same time not to reach it in the neighboring microthrusters, that is, at the border of the computational domain. Additionally, the resistor temperature during the heating phase must not become too

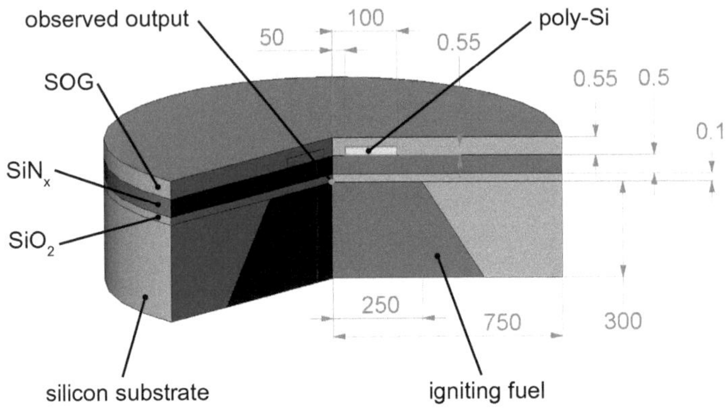

Fig. 5.2. Model structure of 2D axi-symmetrical model for a pyrotechnical microthruster (single igniter wafer). All dimensions are given in μm.

high, as this could cause destruction of the membrane before ignition of the fuel. Subsequent MOR supplies a system-level model which is used for the joint simulation of the of the array and the driving circuitry.

To evaluate MOR algorithms we use a simplified axi-symmetrical model of a single igniter chip (see Figure 5.2), which after finite element based spatial discretization of the governing heat transfer equation (2.5) results in a linear thermal system of 1071 ODEs. The device solid model has been made and meshed in ANSYS with PLANE55 elements. All material properties are considered to be temperature independent. A heat generation rate Q was assumed uniform and applied to the heater. The initial temperature distribution was set to zero and Dirichlet boundary conditions were set to $T = 0°C$ at the bottom of the chip. Numerical simulation results for the full finite element model are shown in Figure 5.3.

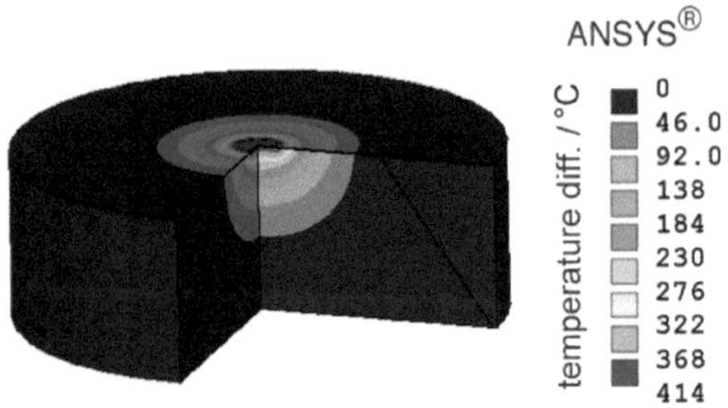

Fig. 5.3. Temperature distribution within the igniter wafer after 0.3 s of heating with 80mW power. $T_{ref} = 0°C$.

5.1.2 Tunable Optical Filter

Most data communication nowadays is performed using fiber-optic cables. The information is carried through fiber-optics by modulated laser light signals which have a fixed wavelength. One application of such communication systems is to connect computer networks to the data 'highway'. This connectivity is realized of various levels, ranging from local area networks to the ultra long-haul installations of fiber-optic cables (more than 10 000km) which connect networks from continent to continent. With ever-increasing amounts of data to be transferred, individually modulated optical channels (with different wavelengths), are being added to the established transmission systems. A major advantage over the electrical data transmission is that there is no crosstalk between light signals. Modern communication systems carry up to 128 independent channels, and systems with 1000 channels on a single fiber are being developed.

With so many channels operating in parallel, network management, for example monitoring of the channel power level and data rate, is becoming very important. Furthermore, individual channels have to be dropped from one fiber and fed into another one. This is usually performed by converting optical signals into the electrical domain and after switching, adding or dropping them, converting them back into the optical domain. This opto–electro–opto conversion is a major limitation on increasing the data rate in fiber-optic cables. In order to circumvent this 'bottleneck', so-called all-optical cross-connects are being developed, where the switching is realized completely in the optical domain. Add-drop functionality is provided by tunable optical filters, which are able to separate the individual channels. Additionally, they serve as key components for optical channel monitoring and tunable lasers.

Several state-of-the-art concepts exist for the realization of tunable optical filters which make use of acousto-optic effects [150, 151] or liquid crystals [152]. As miniaturization of these systems improves their performance considerably (much lower power consumption), a new class of optical MEMS devices is arising. Micromechanically tunable optical filters [153, 154] feature dielectric mirrors in connection with a single air-cavity of variable thickness and achieve tuning by varying the distance between the mirrors. Thermo-optical tunable microfilters [155, 156] on the other hand, are based on "pure" thermal modulation of resonator optical thickness.

The device described here [27], was developed within a DFG project AFON (funded under grant ZA 276/2-1) and represents a new class of tunable optical filter, fabricated using silicon MEMS technology and based on the thermo-optic effect. It consists of an optical resonator which is formed out of two dielectric Bragg mirrors separated by a solid-state material, as shown in Figure 5.4. An important optical characteristics of such a resonator is its optical thickness, defined as $n \cdot d$, where n is the refractive index of the material and d is the distance between the mirrors. The light, which propagates inside the cavity, is reflected repeatedly between both mirrors so that interference occurs. Constructively interfering waves (those with wavelengths $\lambda_0 = 2nd/m, m = 1, 2, \ldots$) may pass through the filter, while others are reflected. The transmission wavelengths are determined only by varying the

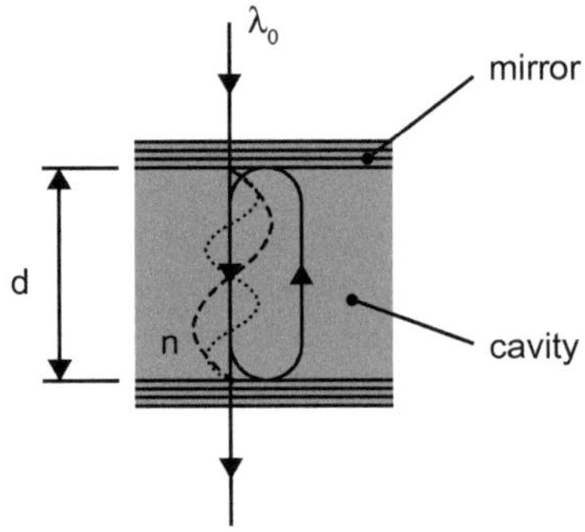

Fig. 5.4. Schema of a solid-state resonator cavity (see also [20]).

temperature and thus the refractive index n of the material, i.e. without physically changing the separation of the mirrors. As the thickness d of the central cavity region (the region between the mirrors) is of the order of the wavelength of light (ca. $1\,\mu\text{m}$), the power consumption (heat the microcavity) is decreased drastically compared to similar devices in [155] and [156].

The device is fabricated by thin film deposition of the optical layers on a silicon substrate and further processing is done using MEMS technology. Amorphous silicon was used for the cavity material because of its suitably high thermo-optic coefficient (change of n with temperature). It is important to mention that changes in the physical dimensions of the cavity due to thermal expansion are negligible and hence have no influence on the shift in transmission wavelength. The Bragg mirrors consist of dielectric materials such as silicon dioxide, silicon nitride or amorphous silicon, which are all deposited by plasma enhanced chemical vapor deposition. Thin film metal resistors are structured on the filter layers for the fabrication of heating and temperature sensing resistors. The microcavity is heated by conducting the electrical current through the heating resistor. In order to reduce the power consumption, the silicon substrate material was removed from the back of the wafer under the filter by anisotropic etching. The removal of substrate material also prevents additional optical resonances. Thermal isolation is further improved by structuring the membrane into a plate and fixing it to the substrate through micromachined suspension arms, as shown in Figure 5.5.

A comprehensive modeling approach for the tunable filter includes the simulation of electro-thermal and thermo-structural domains and the impact of these domains on the optical properties. Although thermally induced structural deformation of the filter membrane has an impact on the optical quality, it is not of primary concern because the overall deformation is limited to buckling of the membrane. The resulting spherical shape has a radius of curvature bigger than 1 mm. As the illuminated area at the center of the membrane is restricted to a diameter of $50\,\mu\text{m}$, the light still passes through the filter almost vertically in this section. Hence, the tuning behavior is mainly determined by the membrane temperature distribution. An inhomogeneous temperature distribution across the illuminated section of the

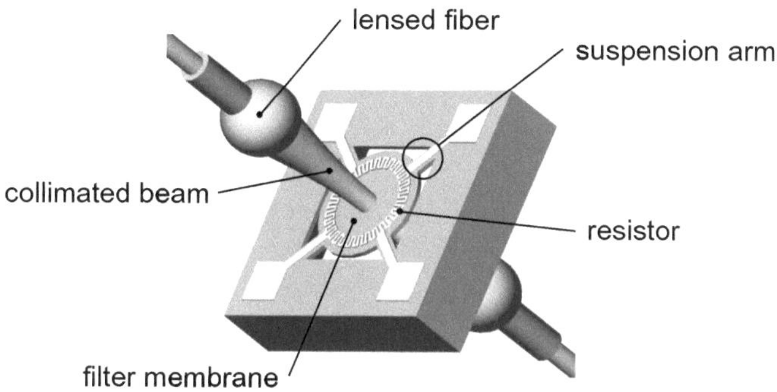

Fig. 5.5. Structure of a tunable optical filter. Picture courtesy of D. Hohlfeld (IMTEK, Germany).

filter membrane causes the filter shape to degrade significantly. Thus, simulation of heat transfer is of paramount concern in the design process of the optical filter. The transient thermal behavior must be carefully analyzed also because the tuning and switching speed are important parameters for future operation of the filter in dynamically configurable optical networks. Beyond the electro-thermal device simulation the connected circuitry can be considered as well. Feedback control can be utilized for stabilization of the membrane temperature. Hence, a system-level simulation becomes necessary and requires a compact thermal model.

To evaluate MOR algorithms we use a two-dimensional device model, which neglects heat transfer in the z direction. Therefore, a weighted sum of material properties of different layers was assigned to corresponding surfaces, as shown in Figure 5.6 All material properties are considered to be temperature independent. A finite element based spatial discretization of the governing heat transfer equation, Eq. 2.5, performed in ANSYS with PLANE55 elements, results in a linear thermal system of 1668 ODEs. A heat generation rate Q was assumed uniform and applied to the heater's surface. The initial temperature distribution is set to zero and Dirichlet boundary conditions $T = 0°\mathrm{C}$ were set at all four edges of the model. The model contains a constant load vector, corresponding to a constant input power of 1 mW. Throughout a single output, placed in the middle of the membrane, will be observed. Numerical simulation results for the full finite element model are shown in Figure 5.7.

5.1.3 Microhotplate Gas Sensor

There is a large demand for gas sensing devices in various domains. They are desired in safety applications where combustible or toxic gases are present or in comfort applications, such as climate controls of buildings and vehicles where good air quality is required. Additionally, gas monitoring is needed in process control and laboratory analytics. All of these applications demand cheap, small and user-friendly gas sensing devices which show high sensitivity, selectivity and stability

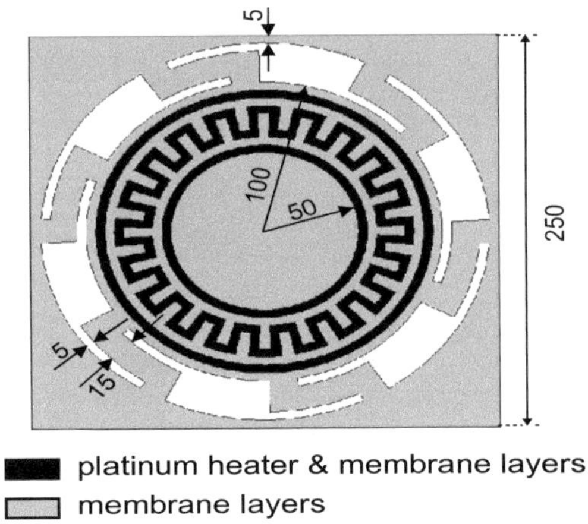

Fig. 5.6. Model structure of 2D model for a tunable optical filter. All dimensions are given in μm.

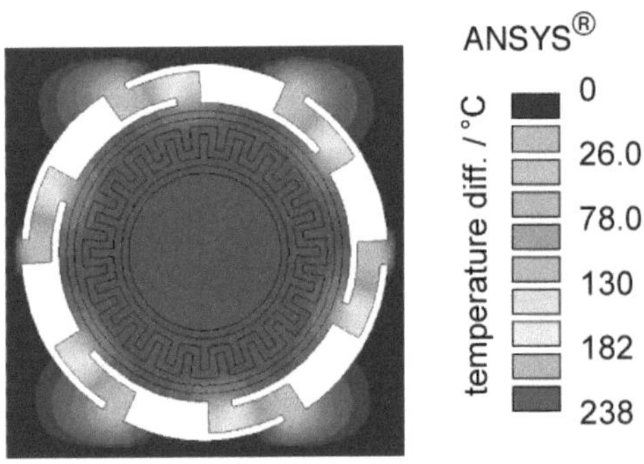

Fig. 5.7. Temperature distribution of the optical filter after 0.25 s of heating at 1 mW. $T_{ref} = 0°C$.

with respect to a given application. A large variety of gas sensors exist based on different sensing principles, semiconductor gas sensors, optical sensors, thermal conductivity sensors, different mass sensitive devices, etc. The main advantages of semiconductor gas sensors based on metal oxides are sensitivity to some relevant gases like CO, H_2, NO_x and hydrocarbons, simple signal processing, low production cost and small size.

In the 1960s it was demonstrated [157] that semiconducting metal oxides heated to $\sim 300°C$ in air vary their conductivity according to the presence of reactive gases in the air. This initiated the commercial development of thick film sensors based mostly on SnO_2. However, SnO_2 gas sensors have low selectivity and high power consumption due to the use of bulky ceramic substrates. Hence, in the last few years a large amount of research has been devoted to the development of low power

Si-micromachined gas sensors based on microhotplates [25, 158, 36]. Usually, the main drawback of such implementations has been the fragility of the thin dielectric membrane used to achieve thermal isolation of the microhotplate from the surrounding wafer.

The goal of the European project Glassgas (founded under IST-99-19003) was to develop a novel microhotplate gas sensor that uses a metal oxide as the sensing material and features low power operation and high mechanical stability. The fabrication approach enables the integration of various sensing areas in an array configuration [161]. The gas-dependent change in electrical conductivity relies upon diffusion of oxygen from an ambient gas (e.g. air) into the sensitive layer at room temperature. In the presence of a reducing or oxidizing gas, which is able to react with the absorbed oxygen at temperatures between 200°C and 400°C , the oxygen surface concentration and thus the electrical conductivity of the metal oxide is altered. This device requires a homogeneous temperature distribution over its gas sensitive regions, good thermal decoupling between the hotplate and the silicon rim and good mechanical stability at high temperatures. This is achieved by supporting the silicon micromachined platform using glass pillars emanating from a glass cap above the silicon wafer, as shown in Fig. 5.8. Glass (Pyrex #7740) was chosen due to its low thermal conductivity and anodic bondability to silicon.

Double-sided polished Si wafers are used for fabrication of the sensor element. A thermal oxide is grown and afterwards photolithography and structuring to define the diffusion area of the later silicon island is carried out. Boron diffusion is performed and subsequently annealing in an oxygen atmosphere. Then the oxide is removed from the front side and a new thin oxide layer is grown prior to deposition of the silicon nitride membrane material. The gas sensor layout features four interdigital structures with symmetrical electrodes. A heater is located on the outer edge of the hotplate areas. Metal electrodes and the heating resistor are fabricated from platinum. Four different gas sensitive layers are deposited and structured in separate steps. After completing the silicon sensor chip, the microstructured glass cap is

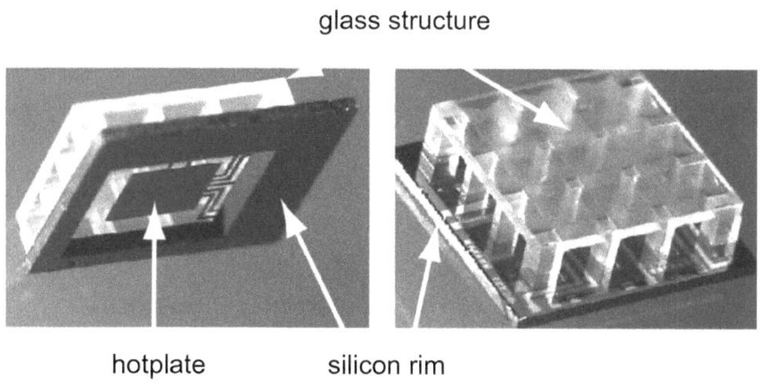

Fig. 5.8. Gas sensor array structure. Picture courtesy of J. Wöllenstein (Fraunhofer IPM, Germany).

bonded to the front side of the wafer. The columns of the supporting glass structure are fabricated by mechanically sawing a 1 mm thick glass substrate. The front-side of the bonded wafer assembly is protected during the following KOH etching step of the silicon back side. Etching stops at the membrane bottom side and at the highly p-doped silicon platform. Subsequent removal of the dielectric membrane material can be performed by dry etching from the back side, so that the silicon island is solely supported by the glass pillars. In this way the hotplate is almost completely thermally isolated from the surrounding silicon substrate.

A micromachined gas sensor is not only a challenge with respect to thermal design but also with respect to mechanical design. Only by choosing the right mechanical design can a large intrinsic or thermal-induced membrane stress leading to membrane deformation/breaking of the membrane be avoided. It is further necessary to build a chemometrics calibration model which correlates the set of sensor resistance measurements to the sensed gas concentration. Prior to fabrication, a thermal simulation is performed to determine the heating efficiency and temperature homogeneity of the gas sensitive regions. Another important thermal issue to be considered by the simulation is the thermal decoupling between hotplate and silicon rim. As the device is connected to circuitry for heating power control and sensing resistor readout, a system-level simulation is also needed. Hence, a compact thermal model must be generated.

We have tested the applicability of Krylov-subspace based MOR to large-scale linear systems using a 3D sensor model shown in Figure 5.9. All material properties are considered to be temperature independent. Temperature is assumed to be in degrees Celsius with an initial state of 0°C and Dirichlet boundary conditions $T = 0°C$ are applied at the top and bottom of the chip. A finite element-based spatial discretization of the governing heat transfer equation (Eq. 2.5) results in a linear thermal system of 73 955 ODEs. The device solid model was made and meshed in ANSYS with SOLID70 elements. The model contains a constant load vector corresponding to a constant input power of 340 mW. Throughout, two outputs for this model for linear and weak nonlinear simulation will be observed. Both are marked in Figure 5.9. Numerical simulation results of the temperature distribution over the chip are shown in Figure 5.10.

5.2 Preparation for Automatic Model Order Reduction

It is presently possible to apply the Arnoldi process only to solely linear or nonlinear input function systems. For electro-thermal models, this implies not only temperature-independent material properties, but also decoupling of the governing equations, Eq. 2.11, Eq. 2.12 and Eq. 2.14. After spatial discretization in ANSYS, the resulting system of coupled domain electro-thermal simulation (assuming constant material properties) has a form:

$$\begin{bmatrix} C_{th} & 0 \\ 0 & 0 \end{bmatrix} \cdot \begin{bmatrix} \dot{T} \\ \dot{\varphi} \end{bmatrix} + \begin{bmatrix} K_{th} & 0 \\ 0 & K_{el} \end{bmatrix} \cdot \begin{bmatrix} T \\ \varphi \end{bmatrix} = \begin{bmatrix} Q(\varphi, t) \\ I(T, t) \end{bmatrix} \tag{5.1}$$

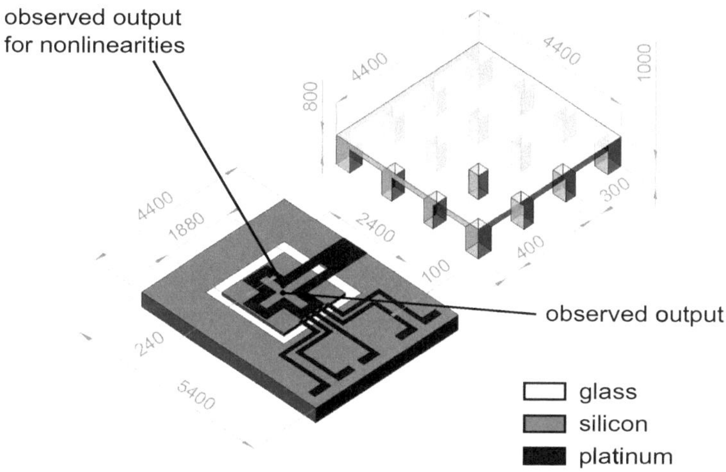

Fig. 5.9. Model structure of a 3D model for a microhotplate gas sensor. All dimensions are given in μm.

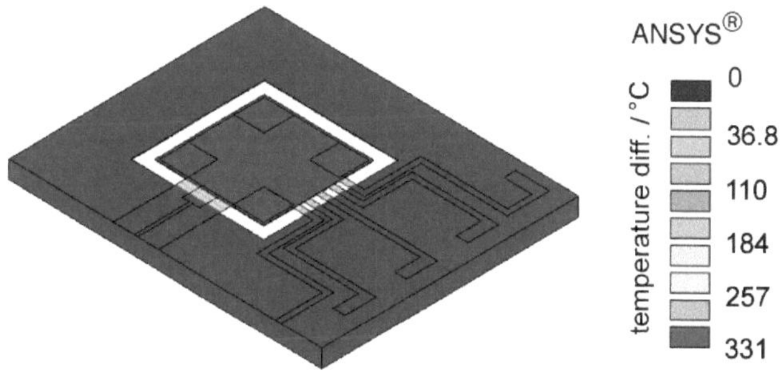

Fig. 5.10. Temperature distribution over the gas sensor chip after 5 s of heating at constant heating power of 340 mW. $T_{ref} = 0°C$.

where C_{th} is a thermal specific heat matrix, K_{th} is a thermal conductivity matrix, K_{el} is electric conductivity matrix, Q is a heat generation vector (in our case, due to Joule heating only), I is a nodal electric current vector, T and φ are the temperature and electric potential vectors. A simplification to the "pure" thermal model can be done by assuming a uniform heat generation rate Q over a lumped heater with electrical resistivity R and assuming that the electric power is completely transformed into heating power. This changes Eq. 2.12 to:

$$Q(t) = I^2(t) \cdot R = \frac{U^2(t)}{R} \tag{5.2}$$

where $I(t)$ is the electric current and $U(t)$ is the applied voltage over a lumped heater. By inserting Eq. 5.2 into Eq. 5.1 we get a linear thermal ODE system, which in the single-input–single-output representation has the form:

$$\begin{aligned} C\dot{T}+KT &= F\cdot Q(t) \\ y &= e^{T}\cdot T \end{aligned} \tag{5.3}$$

where $F, e \in R^{n}$ are the load and the output vectors and n is the dimension of the system.

It turns out that for our case studies it is possible to approximate the temperature distribution of the electro-thermal simulation through the temperature distribution of the thermal simulation. It is done by adding all the Joule heat terms over the heating resistor volume in ANSYS and applying this value as homogeneous heat power over the same heater volume in pure thermal simulation. Figure 5.11 shows the Joule heat distribution over the heater of the gas sensor model, which is proportional to the squared current density distribution. It is approximately uniform. Abrupt changes in the geometry, such as right angles, lead to extreme numerical values (dark blue and red in Fig. 5.11.). They are physically not justified and should be evaluated with caution, as in the fabricated structure the corners are slightly rounded by the photolitography.

Hence, in this case the total amount of heat determined by an electro-thermal simulation can be applied as a homogenous heat generation to the heater's volume. Indeed, Fig. 5.12. and Fig. 5.13., which compare the temperature distribution over the heater for the coupled-domain simulation and for the thermal simulation, show almost identical results. The asymmetric temperature distribution over the heater is due to the asymmetrical model structure, displayed in Fig. 5.11.

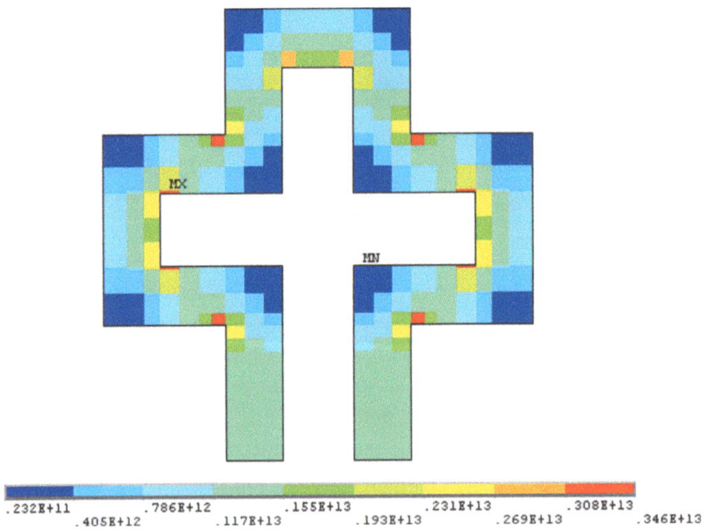

Fig. 5.11. Joule heat distribution in W/m^3 over the heater of the microhotplate gas sensor.

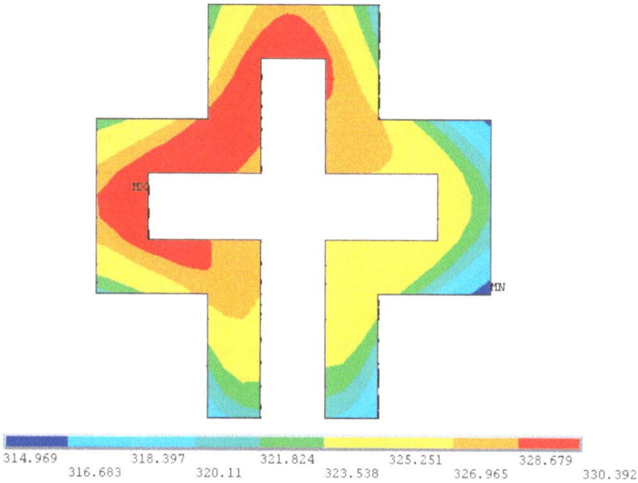

Fig. 5.12. Temperature distribution (in °C) over the gas sensor's heater for the coupled-domain simulation. Electrical voltage $U = 1.77$ V applied to the heater.

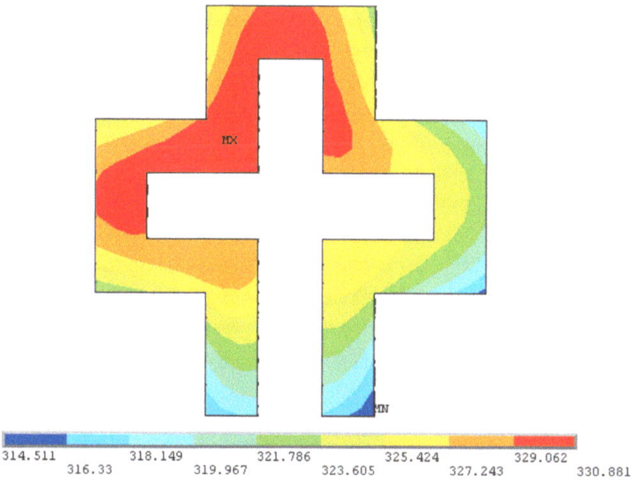

Fig. 5.13. Temperature distribution (in °C) over the gas sensor's heater for the thermal simulation. The sum of Joule heat terms from Fig. 5.11., which is 340mW, was applied as the homogeneous heat power.

We obtained similar results for microthruster and optical filter models as well, which shows that simplification from electro-thermal to pure thermal FE modeling is acceptable for these devices. In general, it is always acceptable when the current density distribution over the heated volume is approximately uniform. This happens when the heater material has a high thermal conductance (such as metal or silicon), and the heater itself is thermally isolated (by means of a membrane for example). In such a case, the high thermal conductance causes an instantaneous distribution of generated heat, which results in an almost homogenous temperature profile.

5.3 Model Reduction of Thermal MEMS via the Arnoldi Algorithm

Let us emphasize once more that the goal of Arnoldi-based MOR is to transform the equation system Eq. 5.3 into a system of the same form:

$$\begin{aligned} C_r\dot{z}+K_rz &= F_rQ(t) \\ y_r &= e_r^T\cdot z \end{aligned} \tag{5.4}$$

but with a much smaller dimension $r \ll n$. Here, a vector of generalized variables z can be seen as a projection of the n-dimensional temperature vector to r-dimensional subspace, subjected to some error ε:

$$T = V\cdot z+\varepsilon, z\in R^r, r\ll n \tag{5.5}$$

and $y_r = e_r^T\cdot z$ is the linear combination of the reduced states, which corresponds to the required states y in Eq. 5.3.

The matrix V (in the previous chapters we also called it V_r) in Eq. 5.5 is composed of r n-dimensional vectors that form a basis for the right Krylov subspace of the dimension r:

$$K_r^R\{A,b\} = span\{b, Ab, \ldots, A^{r-1}b\} \tag{5.6}$$

with $A = -K^{-1}C$, $b = -K^{-1}F$. In fact, the whole Eq. 5.3 can be directly projected onto the reduced subspace by applying Eq. 5.5 (ε is neglected) and then multiplying Eq. 5.3 from the left side by V^T. As already described in Section 3.2, this projection process produces a reduced order system, Eq. 5.4, according to the Padé-type approximation:

$$\begin{aligned} C_r &= V^TCV \\ K_r &= V^TKV \\ F_r &= V^TF \\ e_r &= V^Te \end{aligned} \tag{5.7}$$

It can be shown that the transfer functions of the systems Eq. 5.3 and Eq. 5.4 defined as

$$G(s) = e^T(sC+K)^{-1}F \text{ and} \tag{5.8}$$

$$G_r(s) = e_r^T(sC_r+K_r)^{-1}F_r \tag{5.9}$$

when developed into Taylor expansions around $s_0 = 0$:

$$G(s) = \sum_{i=0}^{\infty} m_i s^i \quad \text{with} \quad m_i = e^T(-K^{-1}C)^i K^{-1} F$$
$$G_r(s) = \sum_{i=0}^{\infty} m_{ri} s^i \quad \text{with} \quad m_{ri} = e_r^T(-K_r^{-1}C_r)^i K_r^{-1} F_r \tag{5.10}$$

where m_i and m_{ri} are called the ith moments, match the first r moments:

$$m_i = m_{ri}, i = 0, \ldots, r-1 \tag{5.11}$$

Note once more that the transformation matrix V is a direct output of the Arnoldi algorithm, and that neither input function $Q(t)$ nor output vector e take part in order reduction. This brings out two additional important properties of the Arnoldi algorithm:

1. The single-input–single-output (SISO) algorithm setup is sufficient to approximate not only a single output response but also the transient thermal response in all finite element nodes of the device. This will be numerically demonstrated in Section 5.3.1.

2. The nonlinearities of the input function can be transferred to the reduced system, i.e. systems with state-dependent input function can be reduced as well. This will be numerically demonstrated in Section 5.3.2.

Let us show, using the case studies, how Arnoldi-based MOR works for transient heat transfer problems in 2D and 3D.

Microthruster: The Arnoldi process was applied to a system of 1071 ordinary differential equations, to iteratively generate several reduced models of different orders. For each reduced model a transfer function $G_r(s)$ over a relevant range of frequencies from 1rad/s to 10^6rad/s was computed. The magnitude of the frequency responses of the full-scale model and reduced models of orders 10, 20 and 50 are shown in Fig. 5.14. As expected, we observe a good match in the frequency domain around the expansion point s_0, which was chosen to be zero. Figure 5.15 shows the step response of the full-scale model in the observed output node (see Fig. 5.2.) for a constant input power of 80 mW and vanishing step response errors (within the initial 0.05 s) for the reduced models of orders 7, 10 and 20. The errors for all three reduced models are less than 0.02% after 0.02 s. The results show that in the case of a microthruster model, it is possible to approximate an ODE system of 1071 equations with only seven equations with a maximum relative error less than 4%.

Optical filter: The Arnoldi process was applied to a system of 1668 ordinary differential equations to iteratively generate several reduced models of different orders. For each reduced model a transfer function $G_r(s)$ over a relevant range of frequencies from 1rad/s to 10^9rad/s was computed. The magnitude of the fre-

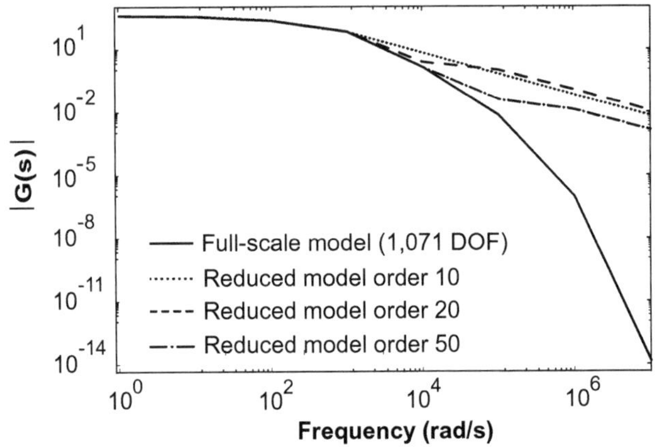

Fig. 5.14. Frequency response of the microthruster.

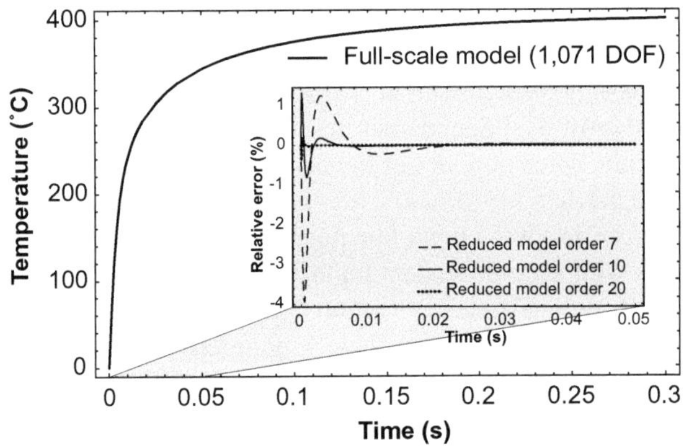

Fig. 5.15. Step response (outer plot) and step response errors (inner plot) of the microthruster for constant input power of 80 mW.

quency responses of the full-scale model and reduced models of orders 1, 5 and 10 are shown in Fig. 5.16. Again we observe an excellent match in the frequency domain around the expansion point $s_0 = 0$. Figure 5.17 shows the step response of the full-scale model in the observed central membrane node (see Fig. 5.6.) for a constant input power of 1 mW and vanishing step response errors (within the initial $0.001\,\text{s}$) for reduced models of orders 1, 5 and 10. The errors for all three reduced models are less than 0.001% after $0.001\,\text{s}$. Hence, in the case of an optical filter model it is possible to approximate an ODE system of 1068 equations with only five equations with a maximum relative error less than 0.01%.

Gas sensor: The Arnoldi process was applied to a system of 73 955 ordinary differential equations to iteratively generate several reduced models of different orders. For each reduced model a transfer function over a relevant range of frequencies from 0.01 Hz to 10 Hz was computed. The magnitude of the frequency

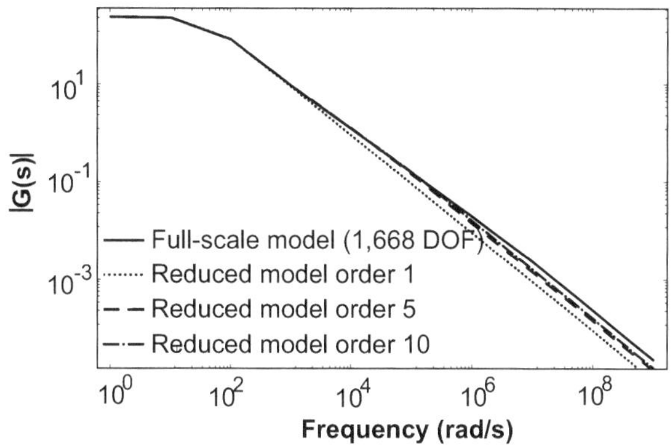

Fig. 5.16. Frequency response of the optical filter.

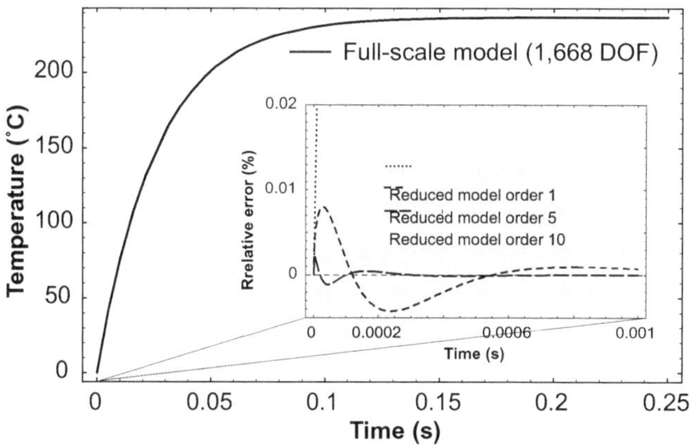

Fig. 5.17. Step response (outer plot) and step response errors (inner plot) of the optical filter for constant input power of 1 mW.

responses of the full-scale model and the reduced models of orders 2, 5, and 10 which show a good match in the frequency domain around the expansion point $s_0 = 0$ are displayed in Fig. 5.18. Figure 5.19 shows the step response of the full-scale model in the observed central membrane node (see Fig. 5.9.) for a constant input power of 340 mW and vanishing step response errors for the reduced models of orders 2, 5 and 10. The errors for all three reduced models are less than 0.5% after 2 s. Hence, in the case of the gas sensor model it is possible to approximate an ODE system of 73 955 equations with only 10 equations with a maximum relative error less than 0.5%.

As expected, the three thermal models show low-pass filter characteristics. This can be proved in general by observing the definition of the thermal system transfer function, Eq. 5.8. A vector term $(sC+K)^{-1}F \in R^{n \times 1}$ consists of n rational functions of the form $P_{n-1}(s)/P_n(s)$, where $P_{n-1}(s)$ and $P_n(s)$ are polynomials in s of the

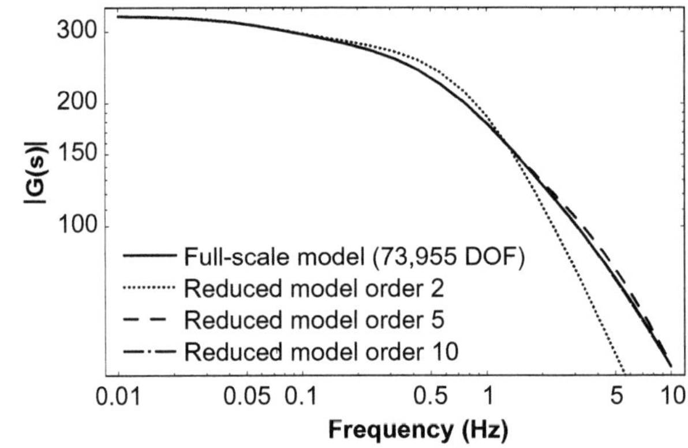

Fig. 5.18. Frequency response of the gas sensor.

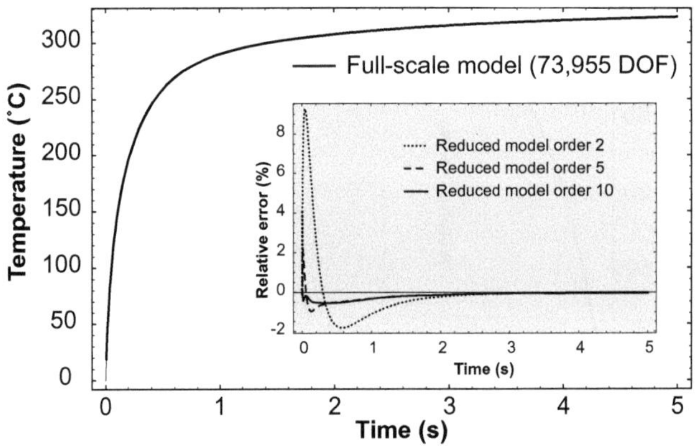

Fig. 5.19. Step response (outer plot) and step response errors (inner plot) of a gas sensor in a central hotplate node, for constant input power of 340 mW.

order $n-1$ and n respectively. This follows from applying the Cramer rule as follows:

$$G(s)=\sum_{i=1}^{n} e_i \cdot \frac{\det\begin{bmatrix} sC_1+K_{11} \ldots F_1 \ldots & K_{1n} \\ & F_k & \\ K_{n1} & \ldots F_n \ldots & sC_n+K_{nn} \end{bmatrix}}{\det\begin{bmatrix} sC_1+K_{11} \ldots & K_{1n} \\ K_{n1} & \ldots sC_n+K_{nn} \end{bmatrix}} \tag{5.12}$$

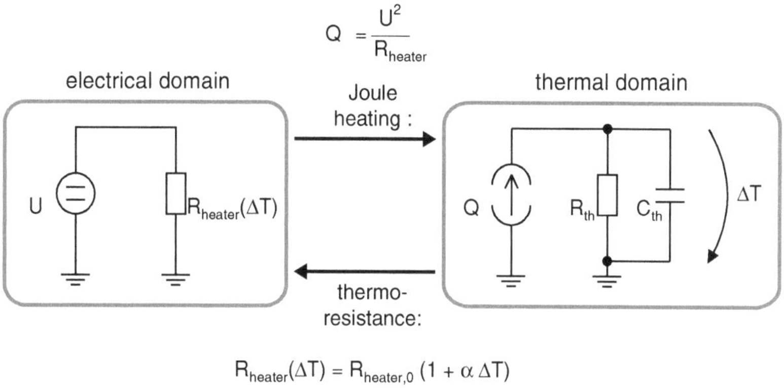

Fig. 5.20. Equivalent thermal circuit corresponding to the low-pass filter characteristic.

From Eq. 5.12 we see that the denominator is a polynomial sum of order $n-1$ since the ith column of $(sC+K)$ is replaced by the constant vector F. The nominator is of order n. Due to the positive diagonal entries of the heat conductivity matrix K, no zeros of $G(s)$ at $s=0$ are present. In order to exclude the vibratory systems with complex conjugated poles, we additionally note that, because C and K are both positive definite (with positive real eigenvalues [185]), the system Eq. 5.3 has only real negative eigenvalues [166], i.e. $G(s)$ is stable with real poles. Hence, all thermal models, described by Eq. 5.3 can be represented by the equivalent thermal circuit shown in Fig. 5.20., which is also a low-pass filter. The approximation in Fig. 5.20. is a first-order approximation, whereas higher order approximations lead to the equivalent thermal networks discussed in Section 2.5.3. In the simplified thermal circuit in Figure 5.20, a thermal resistor R_{th} models heat transfer between the hot areas (e.g. membrane) and the heat sink (e.g. substrate), whereas a thermal capacitor C_{th} models the impossibility of the system to react instantaneously to the heat source Q . This thermal RC element, equivalent to its electrical counterpart, has a time constant $\tau = 1/R_{th}C_{th}$ and therefore the transient thermal response (rise-time) of the device is determined by its thermal mass and its thermal resistance, as expected.

The next important practical issue is that the connection between the rise-time t_e and the cutoff frequency f_g for thermal models can be approximated through the ideal low-pass filter equation:

$$t_e = \frac{1}{2 \cdot f_g} \tag{5.13}$$

This gives the designer guidance on how to choose the necessary frequency range of interest for error estimation (see Chapter 6). For the example electro-thermal case studies the frequency range of interest is between 0 and f_g.

So far we have represented the numerical results either in the frequency domain (Bode diagram–magnitudes of frequency responses) or in the time domain (step responses, errors), which is common practice in electro-thermal engineering [85]. In

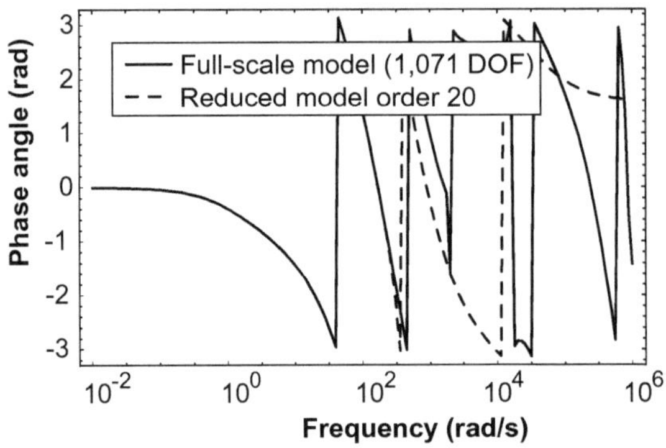

Fig. 5.21. Phase plot of microthruster.

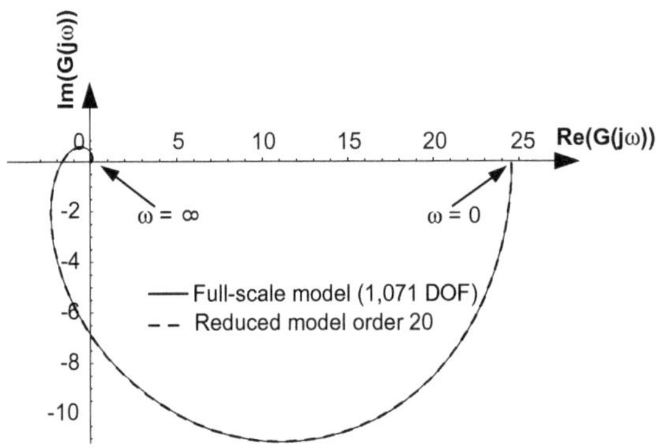

Fig. 5.22. Nyquist diagram of microthruster's frequency response.

principle, there are other representations of dynamic systems, such as pole–zero diagrams and phase plots or Nyquist plots, which originate from control system theory [186]. In Figure 5.21 and Fig. 5.22. we present the phase plot and Nyquist diagram of the microthruster model. The phase angle ranges between $-\pi$ and π , but the plot could actually be unfolded since the phase increases with frequency. Which representation should be used depends on the frequency range of interest, knowledge of the system's stability, etc. A Bode diagram offers a logarithmic representation of the frequency axes, i.e. a huge frequency range can be visualized, as well as a direct correspondence between frequencies and magnitudes. A Nyquist plot represents a frequency response in the complex plane and can be used to judge whether the system is stable (as can the poles–zero plot). However, since we deal with stable (first-order) thermal systems with real poles and use the Arnoldi reduction algorithm, which preserves stability (see Section 3.2.3), the chosen representations suffice for observation of the approximation quality.

5.3.1 Approximation of the Complete Output

The results presented so far refer to the single-input–single-output (SISO) representation of Eq. 5.3, which implies that e and f are both vectors from $R^{n\times 1}$. In more general cases, however, the temperature response in several or even in all finite element nodes may be required. We will refer to the case when the whole temperature field is required as a single-input–complete-output (SICO) setup. In such a case $e = E = I_{n\times n}$ is an identity matrix from $R^{n\times n}$. An important property of the Arnoldi algorithm is that, due to the fact that the output vector (matrix) e does not explicitly participate in MOR, it is possible to recover the complete transient thermal output from the reduced system. In other words, the Padé-type approximation of Eq. 5.8 for the arbitrary vector e, is warranted by only defining a transformation matrix V. Let us explain this. By applying a projection from Eq. 5.5 while neglecting ε, the output row of Eq. 5.3, which is $y = e^T \cdot T$, changes to $y_r = e^T Vz$. Hence, each row of the transformation matrix V "corresponds" to a single node temperature of the FE model. Which output (row) is chosen is defined by e. In the SICO case, all the rows of V are required and

$$y_r = E^T \cdot V \cdot z = V \cdot z \approx T \tag{5.14}$$

stands for the complete temperature vector T.

In order to numerically demonstrate that the projection error ε for thermal models indeed vanishes, let us define a mean square relative difference (MSRD) over time for all the FE nodes as:

$$MSRD(t) = \sqrt{\frac{\sum_{i=1}^{n}\left(\frac{T_i(t) - \hat{T}_i(t)}{T_i(t)}\right)^2}{n}} \tag{5.15}$$

where $T_i(t)$ is the temperature of the ith FE node in time and $\hat{T}_i(t)$ is the ith component of the vector $V \cdot z$ in time.

Figure 5.23 shows a mean square relative difference, which is defined by Eq. 5.15 for the microthruster model computed over the initial 5 ms. We can observe that for the reduced system of order 20, a maximal MSRD for all 1071 nodes amounts to only 0.14%. Hence, after the simulation of the reduced model it is possible to recover the transient solution for all 1071 nodes by using Eq. 5.14.

Figure 5.24 additionally shows the MSRD, Eq. 5.15, for the optical filter over the initial 5ms. In this case, a reduced model of order 5 shows a maximal MSRD for all 1668 nodes less than 0.06%. It decreases even further for higher system orders. As in the previous example, the transient thermal response in all 1668 FE nodes can be extremely well approximated by the reduced models.

As already explained, in both cases, the Arnoldi reduction algorithm can be viewed as a projection from the full space to the reduced Krylov space, Eq. 5.6, with an identity output matrix corresponding to the SICO system description.

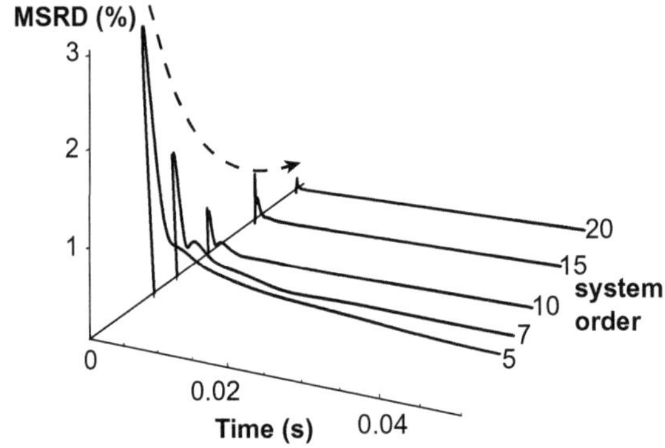

Fig. 5.23. Mean square relative difference (5.12) for all of the 1071 finite element nodes of the microthruster model during the initial 0.05 s.

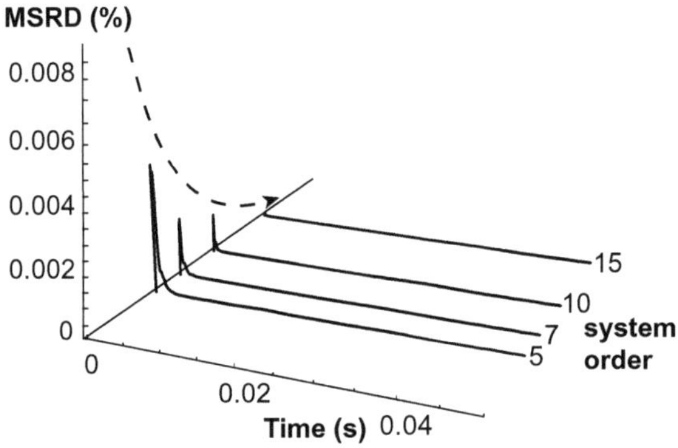

Fig. 5.24. Mean square relative difference (5.12) for all of the 1668 finite element nodes of the optical filter during the initial 0.05 s.

We forego displaying the results of the complete output approximation for the large-scale gas sensor model here, because this is usually not required in engineering applications. Nevertheless, the possibility of approximating the complete transient field with a simple SISO algorithm setup remains an exclusive feature of the Arnoldi reduction algorithm.

5.3.2 Reduction of Systems with Nonlinear Input Functions

When modeling the microthruster, optical filter or gas sensor it is necessary to take into account the dependence of the heater resistivity on temperature, which is given as follows:

$$R = R(T) = R_0 \cdot (1 + \alpha T + \beta T^2 + \ldots) \tag{5.16}$$

where R_0 is resistivity at $0°C$, and α and β are the temperature coefficients. Eq. 5.16 changes Eq. 5.2

$$Q(t,T) = \frac{U^2(t)}{R(T)} \tag{5.17}$$

Thus, a linear equation system Eq. 5.3 changes into a system with a nonlinear input:

$$\begin{gathered} C\dot{T} + KT = FQ(t,T) = F \cdot \frac{U^2(t)}{R(T)} \\ y = e^T \cdot T \end{gathered} \tag{5.18}$$

In terms of the finite element model, the dependence temperature T in $R(T)$, can be chosen as an arbitrary linear combination of states T_i:

$$T = \sum_i c_i \cdot T_i \tag{5.19}$$

It turns out that it is possible to reduce the equation system Eq. 5.18 by using the linear Arnoldi process [167]. As the nonlinear input term $Q(t,T)$, similar to the output vector (matrix) E, is not explicitly affected by the MOR, it is possible to transfer it into the reduced system, which will then be of the form:

$$\begin{gathered} C_r\dot{T}_r + K_rT_r = F_rQ(t,z) \\ y_r = e_r^T \cdot z \end{gathered} \tag{5.20}$$

or, using a projection from Eq. 5.5:

$$\begin{gathered} V^TCV\dot{z} + V^TKVz = V^TFQ(t, V^* \cdot z) \\ y_r = e^TVz \end{gathered} \tag{5.21}$$

The reason why we have written V^* instead of V in Eq. 5.21 lies in Eq. 5.19. In the cases when $i = 1$, heating power is a function of a single node temperature and a product $V^* \cdot z$ defines a single additional output. V^* is then a single row of a projection matrix V. In cases when $i > 1$, V^* is a linear combination of different rows of a projection matrix V. In any cases, $V^* \cdot z$ must be equal to the definition in Eq. 5.19.

We will demonstrate the applicability of Eq. 5.21 numerically for the microhotplate gas sensor example. In order to be able to control heater resistivity, and therefore the temperature of the hotplate, Eq. 5.16 was taken into account by the ANSYS simulation of the full-scale model as well as by MOR. For a platinum sensor heater it is sufficient to assume a linear temperature dependence in the range $0°C$ to $500°C$, i.e. β can be neglected. Measured dependence of the heater's resistivity on temperature and consequently the dependence of the heating power on temperature for a constant input voltage of 14 V are shown in Figure 5.25. A full-scale model was implemented using the nonlinear heat generation rate, which is available as a

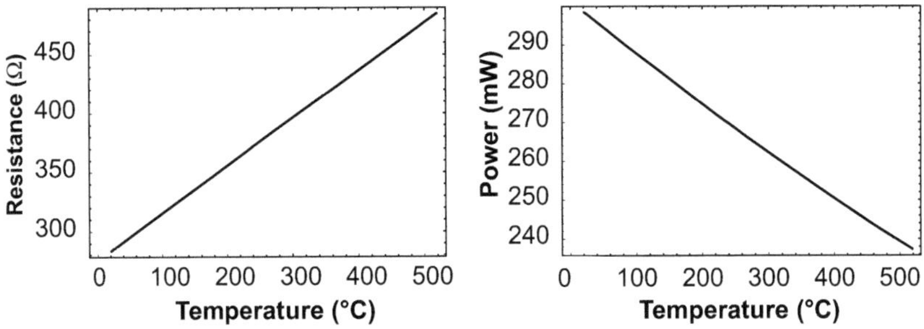

Fig. 5.25. Heater resistivity as a function of temperature (left). Heating power as a function of temperature, measured for constant input voltage of 14 V (right).

table load in ANSYS7. For this procedure, the heater temperature was approximated through a single node temperature T_N (see the observed output for nonlinearities in Figure 5.9), so that the heating power was expressed as

$$Q(t,T) = Q(T_N(t)) = \frac{U^2}{R_0 \cdot (1+\alpha T_N)} \tag{5.22}$$

where U is a desired constant voltage, $R_0 = 274.94\Omega$ and $\alpha = 1.469 \times 10^{-3}$ 1/K . Reduction from order 73 955 down to order 10 was done by linear Arnoldi. As the reduced model contains a constant load vector corresponding to the input power of 340mW, the right-hand side of Eq. 5.20 must be divided by 0.34 before multiplying it with the right-hand side of Eq. 5.22. For integration, T_N must be, of course, replaced with the corresponding linear combination of the reduced states $V^* \cdot z$.

Figure 5.26 shows the step response and step response error between the full-scale and the reduced order 10 model in the observed central hotplate node (see Figure 5.9). Temperature dependent heating power according to Figure 5.25 (right) has been taken into account. These results show that it is possible to approximate the full-scale model of order 73 955 through the reduced model of order 10, with maximum relative error of 1.2%. As expected, the steady-state obtained for the nonlinear sensor model differs from that for the linear thermal model (Figure 5.19), but the reduced models describe these steady-states correctly in both cases.

5.3.3 System-Level Simulation

In our own research, we have used numerical integration in time based on the NDSolve command in Mathematica (see Section 4). For industry-level applications, the reduced models have to be expressed in an appropriate hardware description language (HDL). For example, MAST, a language of the behavioral simulator SABER allows the direct implementation of ODE systems into its templates.

Figure 5.27 shows the structure of the implemented MAST model for a micro-hotplate gas sensor with a temperature-dependent heater as described in Section 5.3.2. It contains a reduced order thermal model with temperature-dependent input

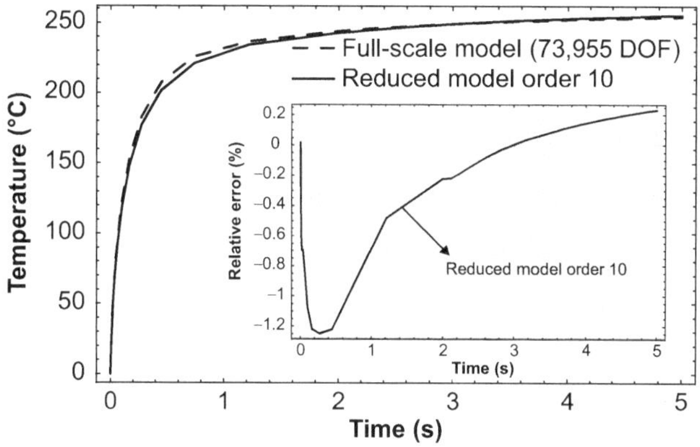

Fig. 5.26. Step response (outer plot) and step response error (inner plot) of the gas sensor in a central hotplate node for temperature dependent heating power according to Fig. 5.12 (right).

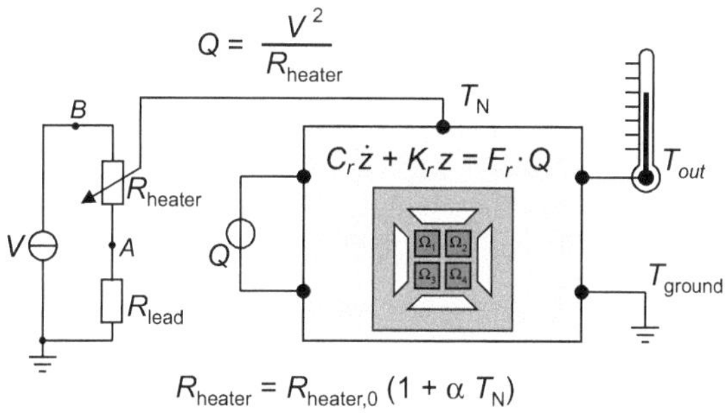

Fig. 5.27. Model structure containing a linear reduced model of the gas sensor with back-coupled temperature-dependent heater.

power (Eq. 5.20). A template containing a (linear) reduced ODE system is back-coupled to the heaters resistivity, which is dependent on a single node temperature T_N (see output node for nonlinearities in Figure 5.9). This temperature is computed as a particular linear combination of the reduced states:

$$T_N = V^* \cdot z \tag{5.23}$$

The output node temperature T_{out} (see the observed output node in Figure 5.9) is also a linear combination of reduced states, marked with y_r in Eq. 5.20. Basically, there is no difference between this implementation and the implementation in Section 5.3.2. Figure 5.28 compares the numerical simulation results of the behavioral MAST model, which was integrated in SABER, the nonlinear full-scale model, which was integrated in ANSYS, and the reduced order model from Eq. 5.20, which was integrated in Mathematica.

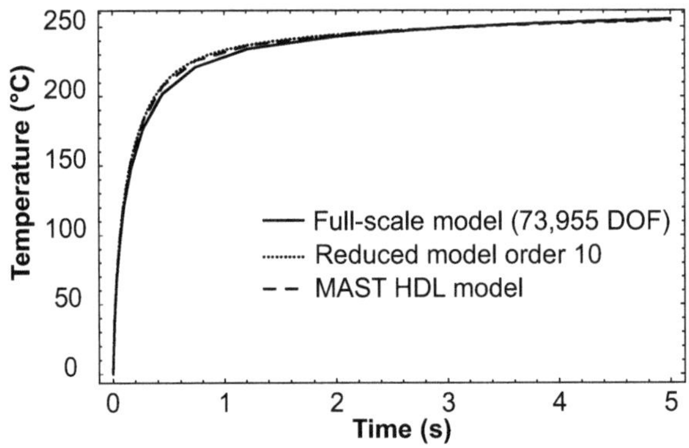

Fig. 5.28. Solution of the full-scale system (73 955 DOF) and of the reduced order 10 system in a central hotplate node of a gas sensor device. The reduced systems were integrated in Mathematica and SABER (MAST HDL model). The temperature dependence of the heating power is based on Figure 5.25.

As expected, the difference between the Mathematica and SABER solutions totals to the numerical integration errors between the different solvers.

An important advantage of the behavioral model over the equation system from Eq. 5.20, however, is the back coupling of the heater resistor. It allows the monitoring of the temperature through the change of resistance and furthermore, certain design changes (such as the change of heater resistivity) are still possible after MOR.

5.3.4 Computational Efficiency

The ultimate goal of MOR is, of course, to increase computational efficiency with minimal loss of accuracy. The high precision of Arnoldi-based reduced order thermal models was numerically presented in the previous sections and here we discuss the decrease in computational time.

In Section 3.2.3, the computational complexity of the Arnoldi algorithm was estimated as

$$2r^2 n + 2rN_z(A) \tag{5.24}$$

where $N_z(A)$ is a number of non-zero elements of a single (preferably sparse) system matrix A [182]. In the case of thermal systems, matrix A is not given explicitly. Instead, the system is described by two large-scale symmetric (sparse) matrices C and K. However, the direct computation of $A = -K^{-1}C$ via matrix inverse would lead to the loss of both, symmetry and sparsity and would require great computational effort. Luckily, the explicit inverse of K can be avoided. If a direct solver is available, either Cholesky or LU sparse factorization of K can be made once:

$$K = L \cdot U \tag{5.25}$$

and then in each iteration of the Arnoldi algorithm, which is:

$$
\begin{aligned}
v_1 &= b = -K^{-1}F \\
v_{i+1} &= K^{-1}C \cdot v_i
\end{aligned}
\tag{5.26}
$$

a fast back substitution (which only takes about 3% of factorization time) for solving Eq. 5.26 is performed using the following three steps:

(1)C is multiplied by v_i, $a = C \cdot v_i$. This is a fast operation, because C is sparse.

(2)The linear equations $L \cdot b = a$ are forward solved, so that $b = L^{-1} \cdot C \cdot v_i$. This is a fast operation, because L is a lower triangular matrix.

(3)The linear equations $U \cdot c = b$ are forward solved, so that $c = (U^{-1} \cdot L^{-1}) \cdot C \cdot v_i$, which is required. Since, U is upper triangular, this is a fast operation as well.

It is worth noting that although the number of nonzero elements of the matrix K – $N_z(K)$ is known, it is not possible to predict $N_z(L)$ and $N_z(U)$ because they depend on the connectivity of K (the distribution of zero entries). In order to speed up the computation and reduce memory storage, K can be reordered (in the sense of moving the nonzeros towards the main diagonal) [166]. Still, $N_z(L)$ is about an order of magnitude larger than $N_z(K)$ and $N_z(K) \geq N_z(C)$ because C is often diagonal. As a result, the steps (2) and (3) of a back substitution require more time than step (1). Hence, our observation is that the MOR of thermal systems via the Arnoldi algorithm can be performed in approximately the same time as a single stationary system solution, that is the time needed for solving $KT = FQ(t)$. Numerical experiments confirm this estimate. The required time totals to the time for factorization and the time for construction of a single Arnoldi vector (via steps (2) and (3)). Unfortunately, it is not possible to estimate the computational effort with more precision since it is impossible to predict $N_z(L)$ and $N_z(U)$ based on $N_z(K)$ only.

Table 5.1 compares the computational times for the stationary solution of the full-scale models and for the construction of the reduced order models for case studies from Chapter 4. In the second and third column the model's dimension and the number of non-zero elements of the matrix K are displayed. The time needed for Arnoldi-based MOR of each model is the sum of the times needed for factorization

Table 5.1. Computational times in seconds on Sun Ultra-80 with 4 Gb RAM and 450 MHz. Times for ANSYS stationary solution includes input and output operation times

	Dimension	Nz(K)	Stationary* solution in ANSYS7.1	Factorization in TAUCS	Generation of the first 50 Arnoldi vectors
Microthruster	1071	5141	3	0.04	0.51
Optical filter	1668	6209	4	0.06	0.69
Gas sensor	73955	885141	200	93	135

and for the generation of the Arnoldi vectors (the last two columns in Table 5.1). Time integration of the reduced, order 50 system in Mathematica lasts about 1 s. The computational times for the transient solutions of the full-scale models in ANSYS with 30 time-steps are 120 s for the microthruster model, 150 s for the optical filter model and 6840 s for the gas sensor model. As the time for the extraction of the system matrices from ANSYS is 317 s for the gas sensor model, the total computational time is reduced by a factor of 12.5 due to MOR.

At the end of this subsection let us note that when the dimension of A grows large enough (e.g. 10^6 DOF), the LU factorization may take too much time. In such a case, Eq. 5.26 must be computed by an iterative method, which can also be based on Krylov-subspace (see [6] and the references there in).

5.4 Arnoldi versus Guyan versus Control Theory Methods

All the algorithms described in Chapter 3 offer the possibility for automatic MOR of the system Eq. 5.3. The mathematically superior control theory methods offer a global error bound as well. Unfortunately, due to the computational effort of $O(n^3)$ for these methods, they are impractical for large-scale systems. The Guyan algorithm can also be used for MOR of large-scale thermal systems, but it requires larger order of the reduced model than Arnoldi to achieve the same accuracy. In this section, Arnoldi-based MOR is numerically compared to MOR using the Guyan algorithm, the singular perturbation approximation, the Hankel norm approximation and the balanced truncation approximation.

Figure 5.29 shows a comparison between the two reduced, order 20 models of the microthruster. One of them was computed by the Arnoldi process and the other by the Guyan algorithm (using ANSYS substructuring). Master degrees of freedom (terminal nodes) needed for the Guyan algorithm were chosen automatically by ANSYS. The maximal relative error using the Arnoldi-based reduction is less than 0.5% (see Figure 5.15), whereas this error by Guyan-based reduced order modeling peaks to over 64% (not shown). This large error by the Guyan algorithm-based reduction mostly occurs during the transient heating phase and vanishes within the steady-state response, according to Eq. 3.45 and Eq. 3.50. The approximation error for the reduced heat capacity matrix Eq. 3.58 decreases as the order r of the reduced system grows. This is demonstrated in Figure 5.30. However, the maximum relative error between the full-scale solution and the reduced solution of order 200 still amounts to 6% (not shown), which is significantly worse than in the case of Arnoldi-based reduction to order 20.

From the microthruster example we can see that the Guyan method offers less accuracy for the reduction of thermal models than Arnoldi. This is because reduced order modeling based on Guyan makes an attempt to generalize Eq. 3.45 for a steady-state thermal response to the transient thermal problem Eq. 5.3 using a coordinate transformation of the form:

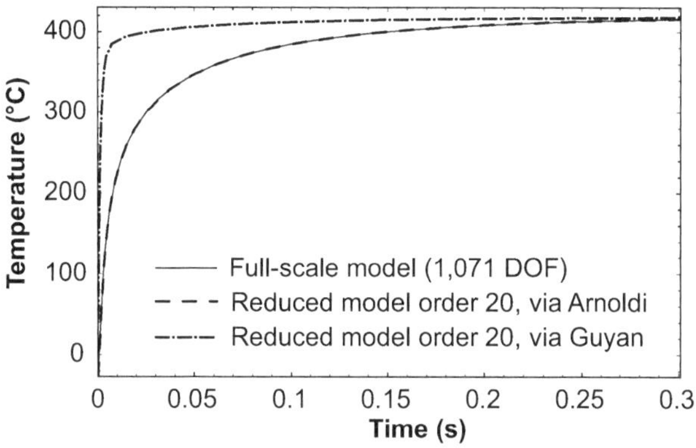

Fig. 5.29. Step response of the microthruster when using the Arnoldi and Guyan algorithms.

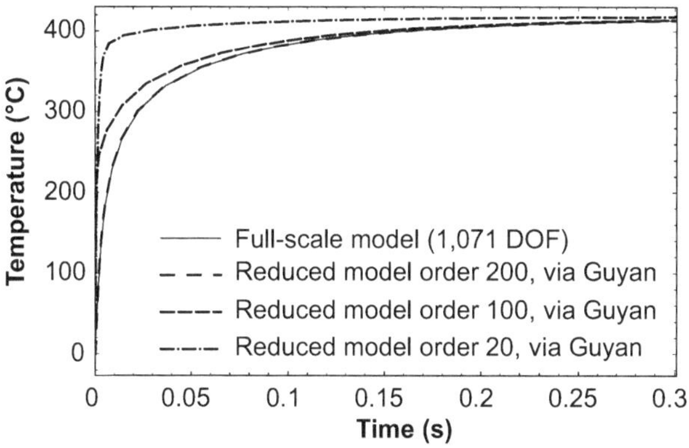

Fig. 5.30. Step responses of the microthruster when using Guyan reduction to different orders.

$$\begin{bmatrix} T_e \\ T_i \end{bmatrix} = \begin{bmatrix} I \\ -K_{ii}^{-1} \cdot K_{ie} \end{bmatrix} \cdot T_e \tag{5.27}$$

This leads to exact matrix condensation of the heat conductivity matrix *K*, but an approximated condensation for the heat capacity matrix *C*.

Arnoldi-based reduction, on the other hand, starts with moment matching for the transient thermal problem Eq. 5.3 as it is, and amounts to a coordinate transformation of the form (5.5), where $V \in R^{n \times r}$ is gained directly as an output of the Arnoldi algorithm and ε vanishes for the case studies described. Another advantage of the Arnoldi algorithm over the Guyan method is its iterative nature. Whereas with the Guyan algorithm Eq. 3.50 and Eq. 3.58 must be recomputed each time the number of master degrees of freedom is changed, only an additional vector matrix multiplication is needed in each iterative step of the Arnoldi orthogonalization procedure. Finally, the fact that no master degrees of freedom need to be chosen by the Arnoldi algorithm contributes to its convenience.

Due to their high computational requirements, the control theory methods from Chapter 3.1 can only be applied to small-scale models with no more than a few thousand degrees of freedom. Hence, we present here the numerical results of MOR of the microthruster model (with 1071 DOF), using the three methods available in the SLICOT library [13]: the balanced truncation approximation (BTA), the singular perturbation approximation (SPA) and the Hankel norm approximation (HNA).

For the specified error bound:

$$\left|G(s)-G_r(s)\right|\leq 0.1 \tag{5.28}$$

a BTA algorithm within SLICOT estimated the size of the reduced system to be $r = 7$. Figure 5.31 shows the relative error of the Arnoldi-based reduction compared to the relative error caused by the BTA-based reduction to the order 7. A maximal relative error by Arnoldi-based reduction approaches 4% during the transient heating phase, whereas in an optimal BTA-based reduction, this error amounts to less than 0.01%. However, the steady-state is better approximated by Arnoldi than by BTA. Figure 5.32 offers a closer look at the relative errors for BTA, SPA and HNA reduction. Out of the three control theory methods, only SPA preserves stationary state. HNA and BTA do not preserve the stationary state, but yield smaller errors within the transient phase than the SPA.

As we have already mentioned, due to their computational complexity the mathematically superior control theory methods are not directly applicable to large-scale FE models. Nevertheless, it is possible to make use of them for such models as well within the framework of sequential MOR. It is based on consecutively applying the Arnoldi algorithm and control-theory methods, and will be numerically demonstrated in Chapter 6.

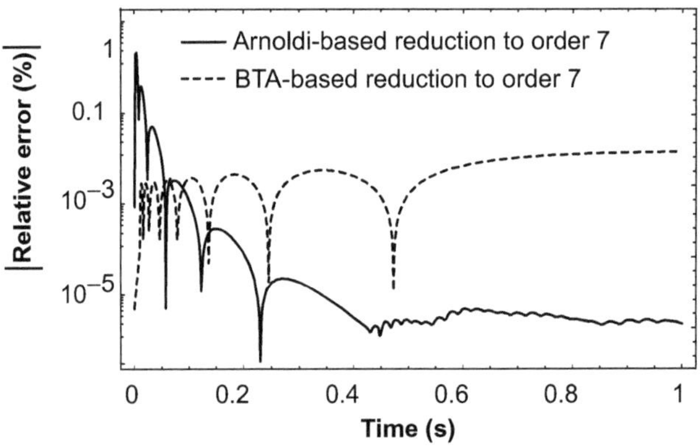

Fig. 5.31. Relative error of BTA and Arnoldi-based reduction for the microthruster model (1071 DOF).

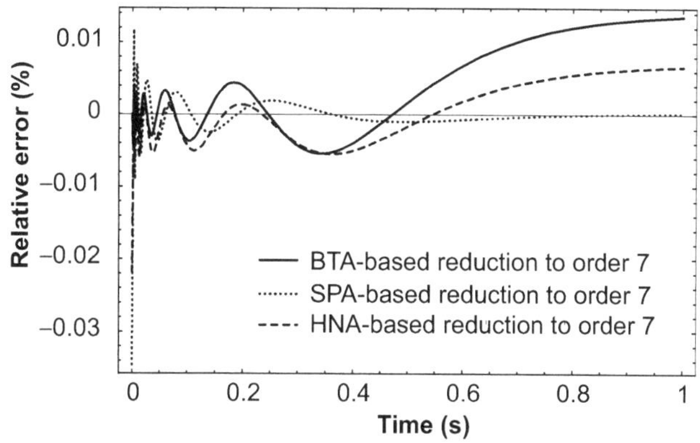

Fig. 5.32. Relative error of BTA, SPA and HNA-based reduction for the microthruster model (1 071 DOF).

5.5 Inverse Thermal Problem via Model Order Reduction

An important engineering task is to build a validated model for each characterized novel MEMS device. It is possible to fit an RC-ladder network to the measured results [162], but in this way we don't obtain a complete physical picture of the device. As in most MEMS applications the whole temperature field has to be known, a more detailed FE model is required. Unfortunately, a common problem here is that the material properties of the thin film materials employed, such as thermal conductivity κ and heat capacity c_p, strongly depend on fabrication conditions and may also be specific for the device under the test. In such a case, it is possible to extract the material properties by fitting a parametrized FE model to a measured transient curve. However, the conventional optimization process is highly time consuming, because in each iteration a time integration of a full-scale model must be performed (see Figure 5.33). As in most MEMS applications the size of an accurate finite element model easily exceeds 100 000 ordinary differential equations (ODEs), we suggest an alternative approach based on MOR. The right path in Figure 5.33 shows that in each iteration of the optimization loop, the suggested approach requires only the time integration of the reduced model (with less than 50 ODEs) and hence brings an enormous saving in computation time. By defining an objective function, which characterizes the difference between simulated and measured results, the data fitting cycle is performed. Below we demonstrate that Arnoldi-based MOR combined with optimization can be used for the efficient extraction of material thermal properties of a microhotplate (see also [164]).

The fabricated structure features a membrane for thermal isolation and two integrated resistors (see Figure 5.34): a thin-film metal resistor, which serves for temperature modulation through Joule heating and a second resistor, which serves as a temperature sensor. The latter is configured for a four-point measurement of the resistance. In order to achieve a preferably circular symmetric and homogenous

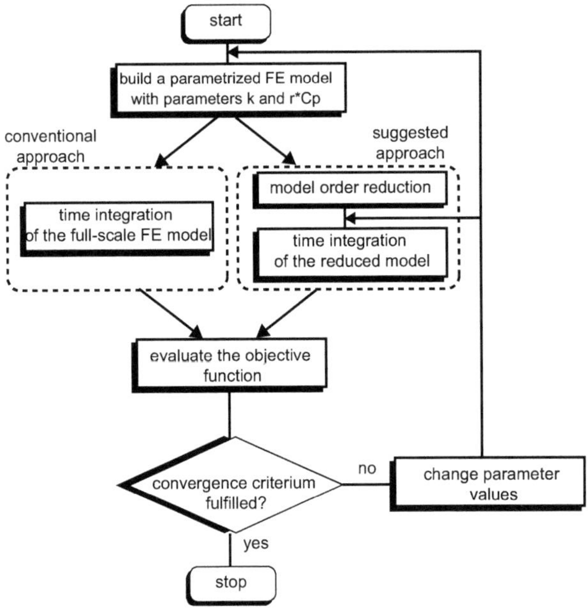

Fig. 5.33. Parameter extraction process for transient thermal problems via optimization and MOR (see also [163]).

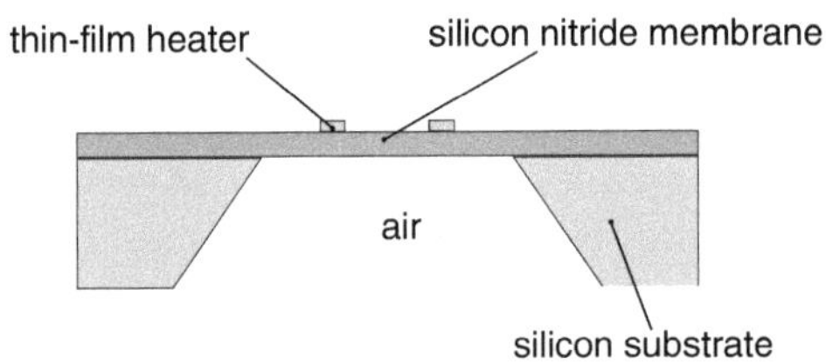

Fig. 5.34. Microstructured silicon/nitride membrane.

temperature distribution at the center of the square membrane, the resistors are arranged as shown in Figure 5.35.

Characterization of the static and transient thermal properties of the filter membrane is performed on a temperature controlled mount. For characterization of dynamic temperature changes, a constant current of 100 μA is passed through the outer terminals of the sensing resistor while the voltage between the inner terminals is measured using an oscilloscope. In order to use the sensing resistor as a temperature sensor, the linear temperature coefficient of the material's resistivity has to be known. The temperature coefficient is measured by acquiring the sensor's resistance at various temperatures. These are precisely set by the Peltier mount. The electrical resistance depends linearly on the temperature over the investigated temperature range. This is modeled as follows:

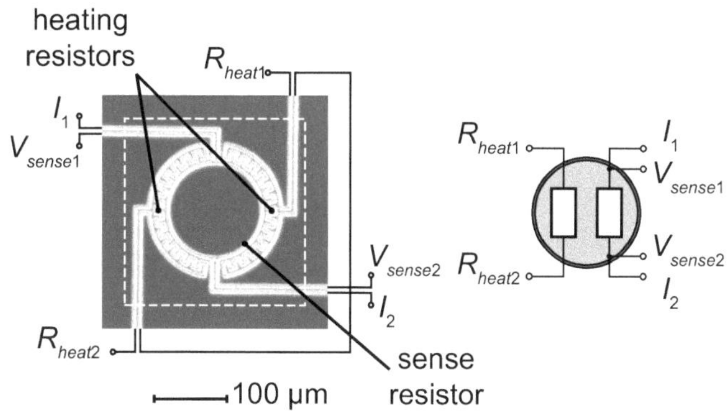

Fig. 5.35. Schematic view of the thin-film resistors for heating and temperature sensing. Heating resistor is operated at constant voltage. The sensing resistor is configured for four-point measurement. Picture courtesy of D. Hohlfeld (IMTEK, Germany).

$$R(\Delta T) = R_0(1 + \alpha \Delta T) \tag{5.29}$$

with R_0 as the resistance for $\Delta T = 0$. A temperature coefficient of $\alpha = 2.293\text{K}^{-1}$ is obtained for a metallization of 150 nm platinum with 50 nm titanium. The electro-thermal dynamics of the system is determined through the specific heat capacity, material density and thermal conductivity of the membrane material. Further influences on transient thermal behavior are the geometry parameters, membrane shape and location of resistors. The transient thermal response of the filter membrane is characterized by applying rectangular heat pulses to the heating resistor using a function generator. The signal output is configured as a voltage source with a fixed output impedance of 50 Ω. After applying the heating power, the membrane temperature increases until a maximum value is reached. This temperature is defined as the steady-state value. After setting the power to zero, the heat stored in the membrane's volume is dissipated to the surrounding media by conduction and free convection. Thus, the temperature drops down to its initial value. The thermal response over a whole period is presented in Figure 5.36.

As our goal is to test the computational environment, we limit ourselves to a simplified two-dimensional finite element model, whose mesh is shown in Figure 5.37. This simplification requires the omission of both out-of-plane thermal conduction and convection from the membrane's surface. The model contains 4402 nodes and Dirichlet boundary conditions $T = 0°\text{C}$ are set at the outer edges of the simulation domain which results in 4182 ODEs.

The layer parameters are differentiated into three sections: areas where only the membrane layers are present, sections with the metal thin-films on the membrane and the silicon frame. The material properties are thickness related values, which are calculated as:

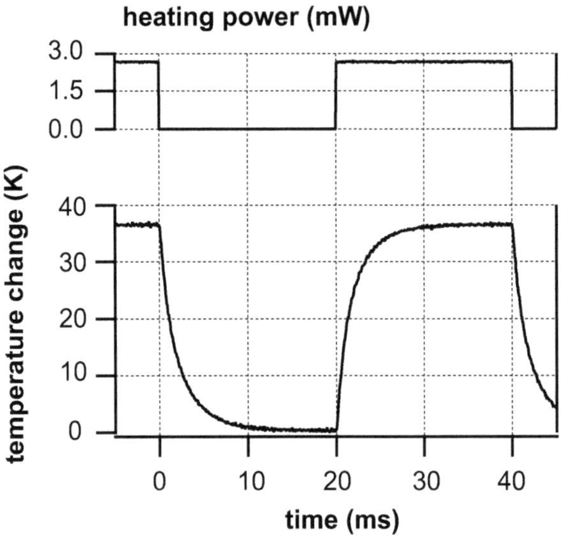

Fig. 5.36. Temperature modulation of a silicon/nitride membrane. A square wave heating power signal with a frequency of 25 Hz was applied to the membrane.

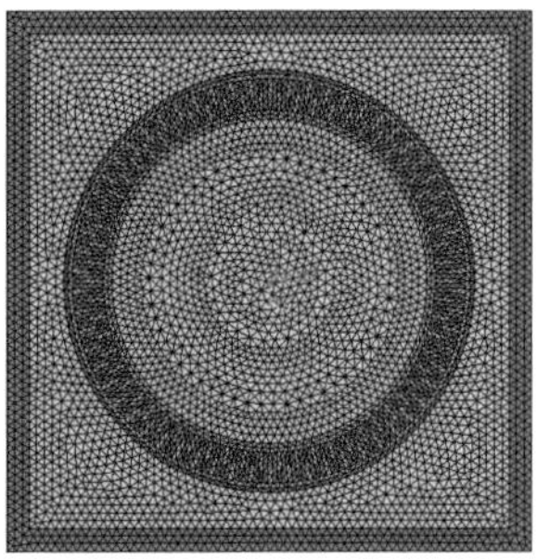

Fig. 5.37. FE mesh of the 2D model with 4402 nodes.

$$\kappa'_m = \sum_i \kappa_i d_i \ , \ \rho'_m = \sum_i \rho_i d_i \ , \ c'_m = \frac{1}{\sum \rho_i d_i} \cdot \sum_i c_i \rho_i d_i \tag{5.30}$$

where d_i is the thickness of each layer and κ_i, ρ_i and c_i are its thermal conductivity, mass density and heat capacity respectively. In order to achieve the consistency between the numerical model and the measurement data, the effective thermal conductivity κ'_m and volumetric heat capacity $c'_m \cdot \rho'_m$ are set as fit parameters in order to determine their actual values via optimization of the reduced model.

Figure 5.38 shows the flexible optimization environment coupled to MOR process. Mathematica is used for scripting, visualization and small size computations. Its function `eval` takes as arguments the fitting parameters κ and $c \cdot \rho$ and calls the external programs ANSYS (for rebuilding the FE model with changed material

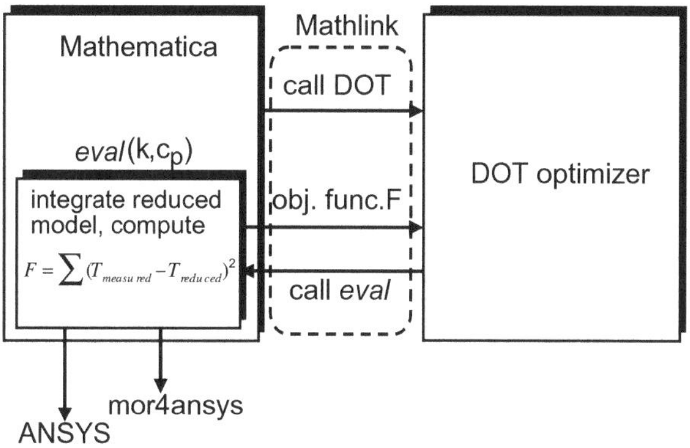

Fig. 5.38. Implementation of the reduced model optimization.

properties) and MOR for ANSYS (for creating a reduced model). It further integrates the reduced model and evaluates the objective function, which is defined as a quadratic error between the measured and the computed curves. Its value is transferred back to DOT optimizer [165] which communicates with *Mathematica* via *MathLink* (our implementation can be found at *http://evgenii.rudnyi.ru/soft/dot/*).

Figure 5.39 shows the measured temperature response and the simulated temperature response for the reduced order 10 model before the optimization (initial values for the material parameters were chosen as $\kappa = 6.35 W/m \cdot K$ and $\rho \cdot c_p = 525 \cdot 10^3 J/(m^3 \cdot K)$). Figure 5.40 shows the measured and the simulated temperature response after 35 cycles of optimization (end values for the material parameters were $\kappa = 5 W/m \cdot K$ and $\rho \cdot c_p = 662 \cdot 10^3 J/(m^3 \cdot K)$). Due to the fact that we have used a simplified 2D model under neglecting out-of-plane thermal conduction and convection and that we have performed optimization with only two parameters, the achieved numerical values should be considered as effective values only. Nevertheless, we have demonstrated that such an approach works in principle.

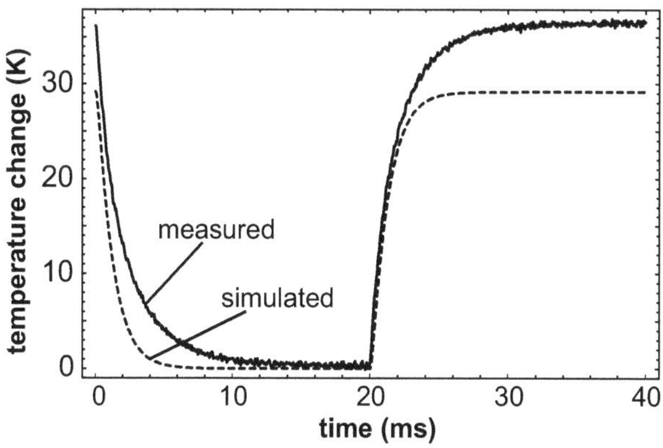

Fig. 5.39. Measured and simulated temperature response before optimization.

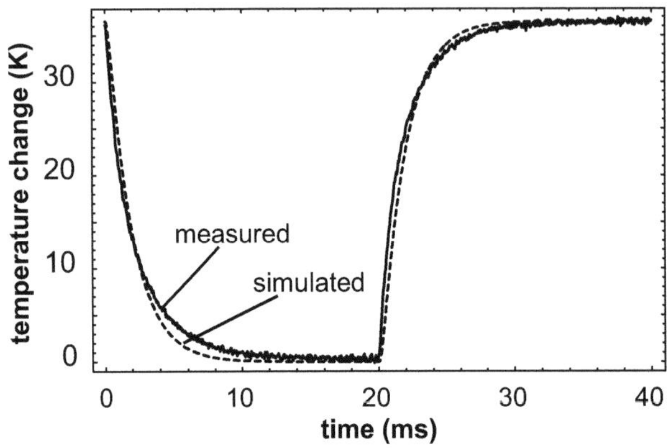

Fig. 5.40. Measured and simulated temperature response after 35 iterations during optimization.

5.6 Summary

In this chapter we have numerically demonstrated that, although the Arnoldi algorithm is presently limited to linear systems and has no global error estimate, it can already be used for highly effective modeling and simulation of electro-thermal MEMS devices. It appears to work extremely well for heat transfer in 2D and 3D, which is consistent with the recent observations of other groups [84, 85] which have been published during our work.

Table 5.2. Main characteristics of Arnoldi algorithm when applied to electro-thermal MEMS models

Arnoldi for electro-thermal models
approximation of the complete output
reduction of weakly nonlinear systems
increase in calculation efficiency
formal conversion into HDL
more accurate than Guyan
more efficient than control theory

The Arnoldi algorithm allows restoration of the whole temperature domain, and can thus be viewed as a fast time integration procedure. It allows the reduction of weakly nonlinear systems, for the case when the nonlinearity appears in the input function. The computational complexity of the reduction of thermal models via Arnoldi is comparable to a single stationary system solution. For the same accuracy the Arnoldi algorithm requires a smaller system order than the commercially available Guyan algorithm does for the thermal systems. It is much more efficient than the mathematically superior control theory methods. The major characteristics of the Arnoldi-based reduction of electro-thermal models are summarized in Table 5.2.

6

Advanced Development

In the previous chapters we have introduced a number of methods for MOR of linear systems and have numerically demonstrated how they work for selected electro-thermal MEMS models. We have also presented software that allows us to use it in practical work right now. However, there are still unsolved problems and the goal of this chapter is to discuss the advanced development in model reduction.

We start with a problem to choose the dimension of a reduced system in Sections 6.1. Model reduction based on Padé approximation does not have the global error estimate and this leaves a practitioner with a question on when to stop the Arnoldi process. We consider several strategies that allow us to overcome this problem in practice.

In Sections 6.2, we address a problem of coupling reduced models between each other. After all, the goal of compact modeling is to make a multiport low-dimensional representation of the device in question. However, there is a big difference between electrical circuits and MEMS devices and we discuss possible ways to cope with this problem.

Model reduction is a rapidly developing area and we finish our book by briefly reviewing the latest research papers on model reduction in Section 6.4. We start with a new development that allows one to preserve several parameters in system matrices during model reduction in the symbolic form. Then we discuss different options for model reduction of second-order systems. After that, we finish by reviewing the latest development for nonlinear model reduction.

6.1 Error Indicators for Heat-transfer Macromodeling of MEMS

In order to apply Arnoldi-based MOR, the MEMS designer has to provide a discretized model (e.g. a finite element (FE) model) of the device and specify which frequency band should be well approximated by the compact model. This is done by choosing one or more expansion points in the frequency domain. The next important step is to specify the desired order of the target reduced system. A key question

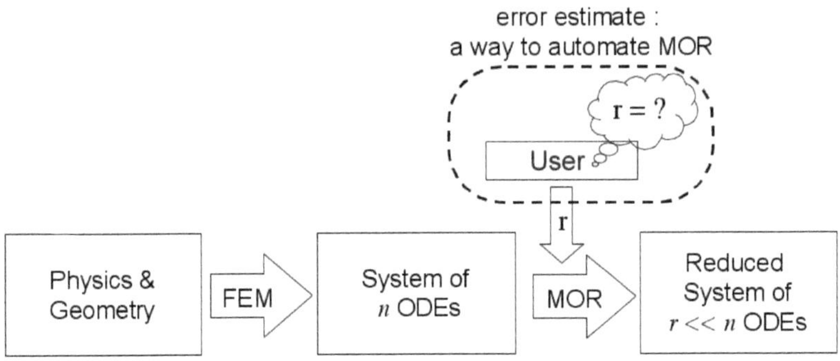

Fig. 6.1. Compact model extraction. Eliminating the need for user iteration makes the process fully automatic.

is: what order of reduced system do we need to select to achieve a desired accuracy. A reduced model is an approximation of the original large-scale model. Hence, the difference between the two can be characterized by some error norm. To automate the MOR process completely, one should be able to estimate this error as a function of the reduced model's dimension. The automatic procedure from device-level to system-level modeling is shown schematically in Figure 6.1.

Unfortunately, an effective error estimate for the Krylov-subspace methods is still an open research question. To our knowledge, only local (single-frequency) error estimates have been suggested so far [168–170]. Recent suggestions for the Arnoldi reduction algorithm can be found in [171] and [172]. The estimate [168] has been developed for the Padé approximation of transfer functions via the Lanczos process and is based on the properties of the tridiagonal transformation matrix created by this process. Hence, it cannot be applied directly to Arnoldi-based reduction. Estimates in [169] and [170] are based on residuals of the associated linear system involved in the transfer function. Evaluation of these general expressions, as well as the choice of the right frequency domain, remains a difficult task. A suggestion in [171] involves numerical error accumulation during the Gramm–Schmidt orthogonalization procedure, which leads to linear dependence of the rth basic vector from the previous r-1 and can be checked by either checking the vector norm or the angle between the vectors. However, this stop criteria offers no relation to the actual approximation error between the original and the reduced order model. Furthermore, numerical error accumulation by computing the projection matrix V may become significant only after a large number of iterations (several hundred). For engineering applications, an error estimate should be implemented in the design flow in such a way as to fit in the iterative framework of Krylov-based MOR.

In this section, we propose three "heuristic" approaches for estimating the error of the reduced-order model computed via the Arnoldi algorithm. The idea is either to compute the relative error between the successive reduced order models (similar to [170]) or alternatively, to compute the Hankel singular values of the reduced model in each iteration of the Arnoldi algorithm. The third approach is based on sequential strategies [173, 174]. Since we cannot yet prove rigorous error bounds, we refer to our results as error indicators. The proposed strategies can be used by MEMS designers to automatically create heat-transfer macromodels using the Arnoldi algorithm.

6.1.1 Convergence of Relative Error

Probably the simplest approach to estimate the model error in either the time or frequency domain is to compute the difference between two "neighboring" reduced models with order r and $r+1$. Let us define a relative frequency-response error as

$$E_r(s) = \frac{|G(s)-G_r(s)|}{|G(s)|} \tag{6.1}$$

where $G(s)$ and $G_r(s)$ are the transfer functions of the original and of the reduced order model (Equation 5.8 and Equation 5.9), respectively. Let us further define a relative frequency-response error between two successive reduced order models as

$$\hat{E}_r(s) = \frac{|G_r(s)-G_{r+1}(s)|}{|G_r(s)|} \tag{6.2}$$

We have observed that for all the case studies described in Chapter 4:

$$E_r(s) \approx \hat{E}_r(s) \tag{6.3}$$

holds for a wide range of frequencies around the expansion point $s_0 = 0$.

As already mentioned, this error indicator can be applied in the time domain as well. Let us define a quadratic relative step-response error as

$$\varepsilon(r) = \frac{1}{N} \cdot \sqrt{\sum_{t_i=0}^{N\cdot\Delta t} \left(\frac{y(t_i)-y_r(t_i)}{y(t_i)}\right)^2} \tag{6.4}$$

where $y(t_i)$ and $y_r(t_i)$ are the system outputs of the full and order r reduced system in N discrete time-points spaced Δt apart. Let us further define a quadratic relative step-response error between two successive reduced order models as

$$\hat{\varepsilon}(r) = \frac{1}{N} \cdot \sqrt{\sum_{t_i=0}^{N\cdot\Delta t} \left(\frac{y_r(t_i)-y_{r+1}(t_i)}{y_r(t_i)}\right)^2} \tag{6.5}$$

Again, it can be shown that for the electrothermal MEMS models from Chapter 4:

$$\varepsilon(r) \approx \hat{\varepsilon}(r) \tag{6.6}$$

Note that expressions in Equation 6.4 and Equation 6.5 are not functions of time, but rather of the system order, and that they require a time integration of system outputs for a chosen output node. Hence, they are slightly more expensive to compute than the frequency-response errors from Equation 6.1 and Equation 6.2 and are susceptible to error accumulation. Equation 6.3 and Equation 6.6 are the main results

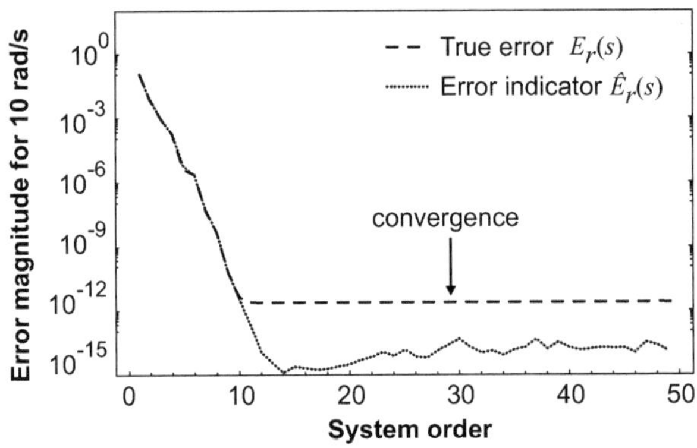

Fig. 6.2. Error indicator in the frequency domain for the microthruster model (1071 DOF) at $\omega = 10\text{rad/s}$.

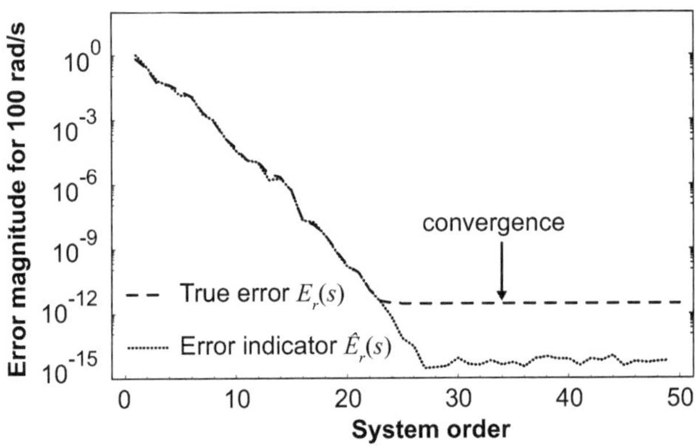

Fig. 6.3. Error indicator in the frequency domain for the microthruster model (1071 DOF) at $\omega = 100\text{rad/s}$.

of this section, and are demonstrated numerically using MEMS case studies described in Chapter 5.

Figure 6.2 through Figure 6.6 compare the true error $E_r(s)$ with the error indicator $\hat{E}_r(s)$ for the microthruster model (1071 DOF) at different frequencies.

Both curves E_r and $\hat{E}_r$ match well in the frequency domain around the expansion point (which is zero in our case). Additionally we can observe two effects:

At frequencies up to $\omega = 10^3\text{rad/s}$ convergence occurs when a threshold reduced system is reached (Figure 6.2 and Figure 6.3). This means that for $\omega = 10\text{rad/s}$ it is not possible to approximate the system better with more than 10 Arnoldi vectors. The system order necessary to reach convergence increases at higher frequencies (see 3D representation in Figure 6.5), since always more terms are needed for accurate Taylor series expansion. The convergence occurs presumably because the machine's numerical precision has been reached. The minimal

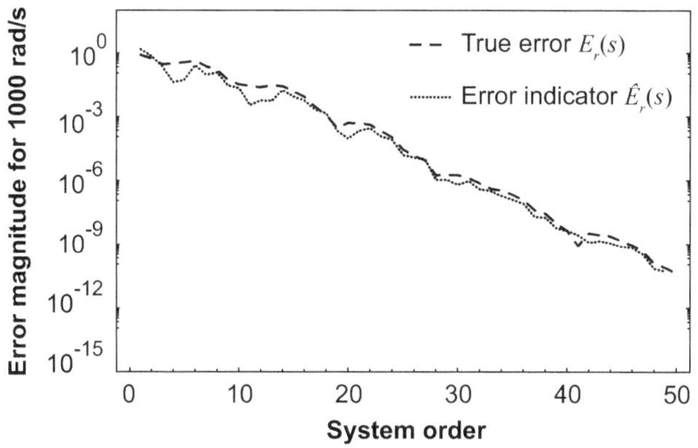

Fig. 6.4. Error indicator in the frequency domain for the microthruster model (1071 DOF) at $\omega = 1000\text{rad/s}$.

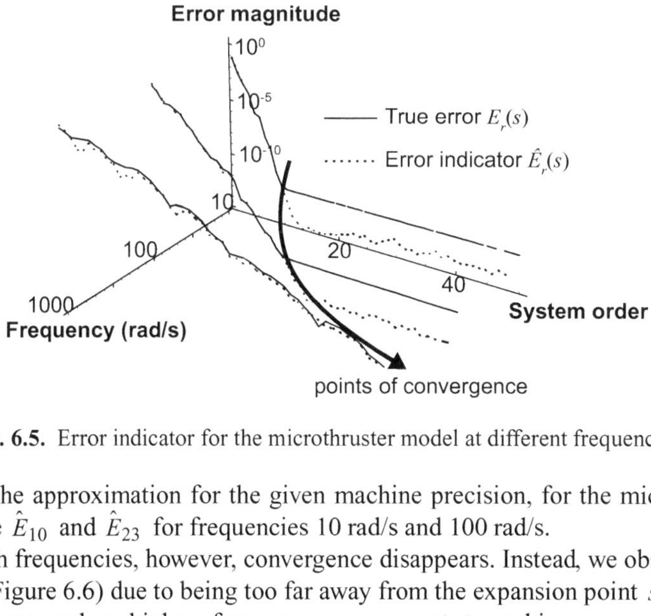

Fig. 6.5. Error indicator for the microthruster model at different frequencies.

errors of the approximation for the given machine precision, for the microthruster model, are $\hat{E}_{10}$ and $\hat{E}_{23}$ for frequencies 10 rad/s and 100 rad/s.

At high frequencies, however, convergence disappears. Instead, we observe fluctuations (Figure 6.6) due to being too far away from the expansion point $s_0 = 0$. For expansion around a higher frequency, we expect to achieve convergence at $\omega = 10^5 \text{rad/s}$ as well.

Figure 6.7 through Figure 6.9 show a good match between the true error $E_r(s)$ and the error indicator $\hat{E}_r(s)$ for the optical filter model (1668 DOF) at different frequencies.

The minimum errors of the approximation for the given machine's precision, for the optical filter model, are $\hat{E}_6$, $\hat{E}_{16}$ and $\hat{E}_{80}$ for frequencies 10, 10^3 and 10^4 rad/s respectively. Due to the slow change of magnitude of $G(s)$ (only five orders of magnitude between 1 rad/s and 10^6 rad/s) in the case of the optical filter,

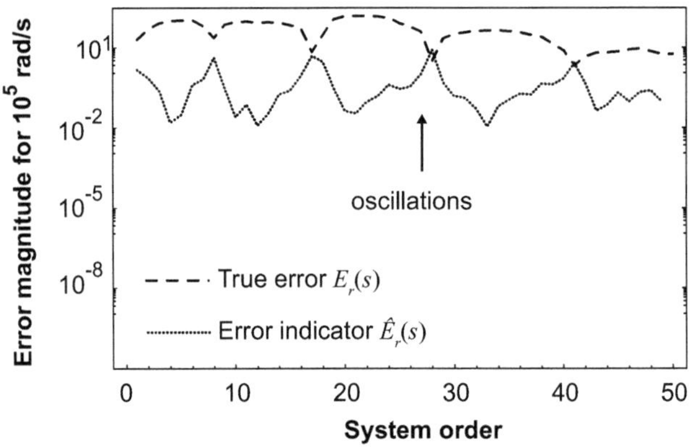

Fig. 6.6. Error indicator in the frequency domain for the microthruster model (1071 DOF) at $\omega = 10^5$ rad/s.

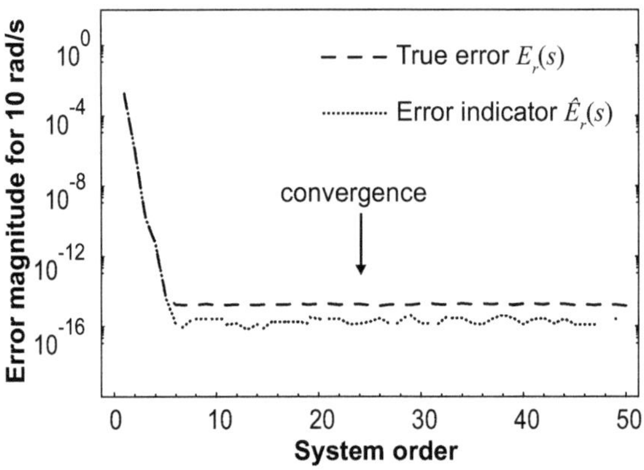

Fig. 6.7. Error indicator in the frequency domain for the optical filter model (1668 DOF) at $\omega = 10 rad/s$.

convergence is still reached at relatively high frequencies. Yet, for higher frequencies, fluctuations would probably appear again and an expansion around $s_0>0$ would become necessary.

Figure 6.10 and Figure 6.11 compare the true error $E_r(s)$ with the estimate $\hat{E}_r(s)$ for the large-scale model of the gas sensor (73 955 DOF) at different frequencies. The results are of the same quality as for the 2D models in spite of the tremendous difference in the model dimensions. Due to the relatively small range of observed frequencies (not far away from the expansion point), no fluctuations occur.

Let us now show the numerical results for the time domain error indicators according to Equation 6.5 and Equation 6.6. Figure 6.12 shows a good match between the true error in the time-domain $\varepsilon(r)$ and the estimate $\hat{\varepsilon}(r)$ for the microthruster model, computed for discrete times between 0 s and 0.3 s with $\Delta t = 0.01$ s.

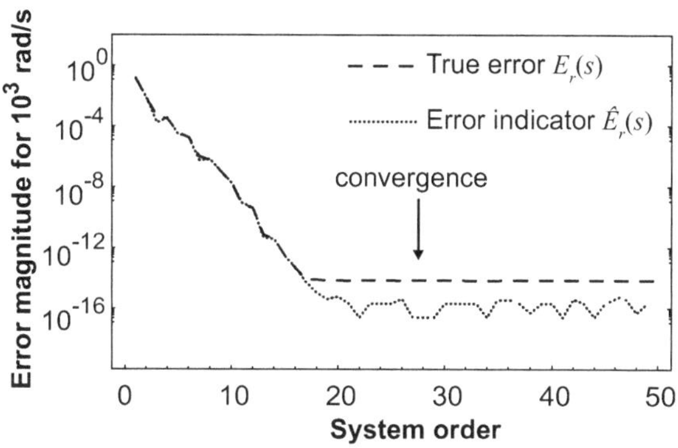

Fig. 6.8. Error indicator in the frequency domain for the optical filter model (1668 DOF) at $\omega = 10^3$ rad/s.

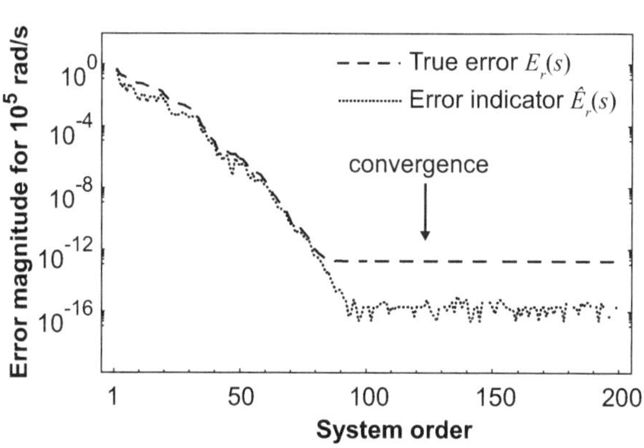

Fig. 6.9. Error indicator in the frequency domain for the optical filter model (1668 DOF) at $\omega = 10^5$ rad/s.

The time interval for error estimates was chosen based on the rise time of the microthruster device, which is approximately 0.3 s (see Figure 5.15). Due to the fact that the Arnoldi algorithm preserves a stationary-state for the expansion around zero frequency, the main error accumulation can be expected during the rise time of the device and therefore the chosen time interval must be long enough. The convergence of time-domain error (after the order 25 of the reduced system has been reached) is in this case also due to the limited numerical precision of the time integrator in Mathematica. `NDSolve` function in *Mathematica* comprises six significant digits. It is possible to increase the numerical precision of time-integration, however, this leads to the common trade-off between efficiency and accuracy.

Figure 6.13 compares the true error in time-domain $\varepsilon(r)$ with the estimate $\hat{\varepsilon}(r)$ for the optical filter model, computed for discrete times between 0 s and 0.25 s with $\Delta t = 0.01\,\text{s}$. The time interval for the error estimate was again chosen based on the rise time of the device (see Figure 6.4).

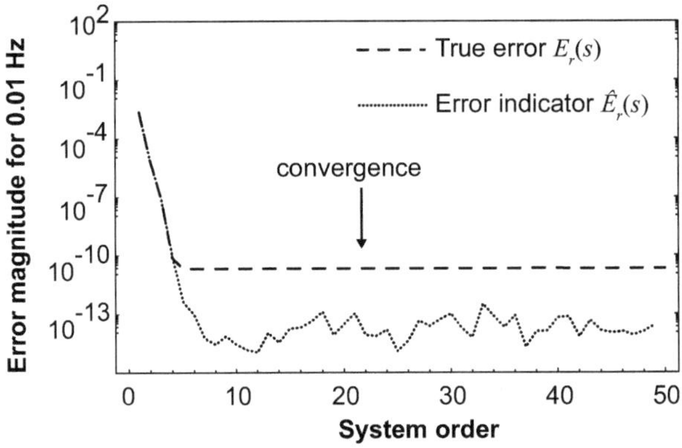

Fig. 6.10. Error indicator in the frequency domain for the gas sensor model (73955 DOF) at $f = 0.01$Hz.

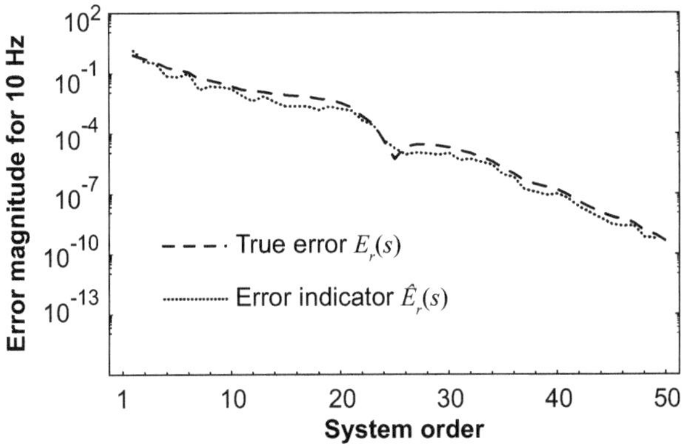

Fig. 6.11. Error indicator in the frequency domain for the gas sensor model (73955 DOF) at $f = 10$Hz.

The time domain estimate of the gas sensor model (Figure 6.13) shows that the integration in ANSYS6.1 with 50 time-steps results an error of about 1%. This can be corrected by increasing the number of integration time-steps within ANSYS. With 200 time-steps an improvement of about 8% (the difference between both true errors are computed as in Equation 6.4) appears. Eventually, the true error and error estimate fit together. Note that in this case the reduced solution is in a way "better" than the full-scale solution of the ODE system computed by ANSYS6.1 due to switching to a more accurate integrator in Mathematica4.1. Hence, in order to circumvent numerical errors caused by additional time integration, it is simpler to use a frequency domain estimate for which only the linear equation systems (in the Laplace domain) have to be solved. The strategy presented in the frequency domain is schematically summed up as an algorithm in Figure 6.15.

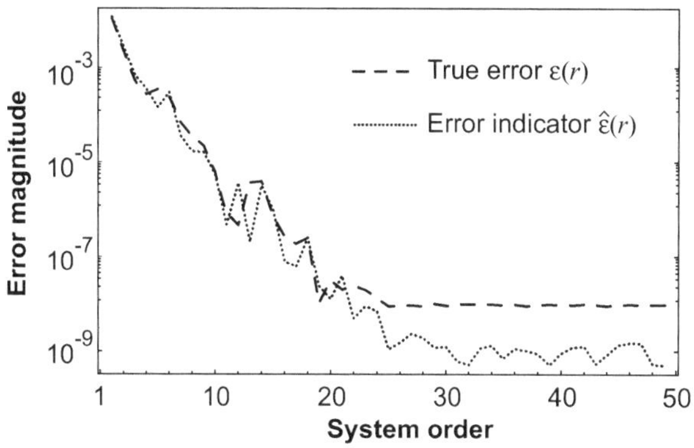

Fig. 6.12. Time domain estimate of the microthruster model (1071 DOF).

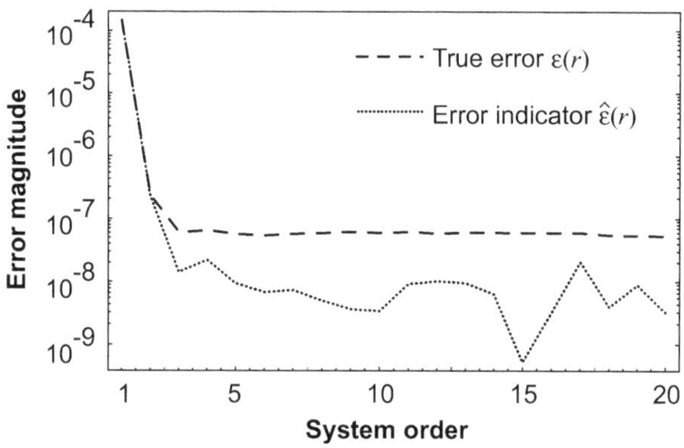

Fig. 6.13. Time domain estimate of the optical filter model (1668 DOF).

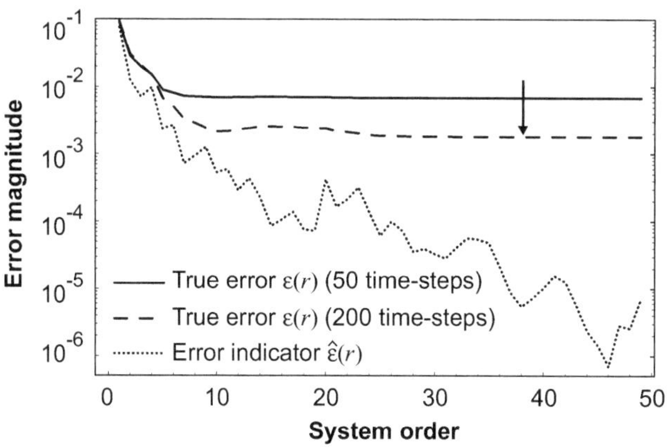

Fig. 6.14. Time-domain estimate of the gas sensor model (73955 DOF). The arrow indicates that the true error decreases as the integration precision increases.

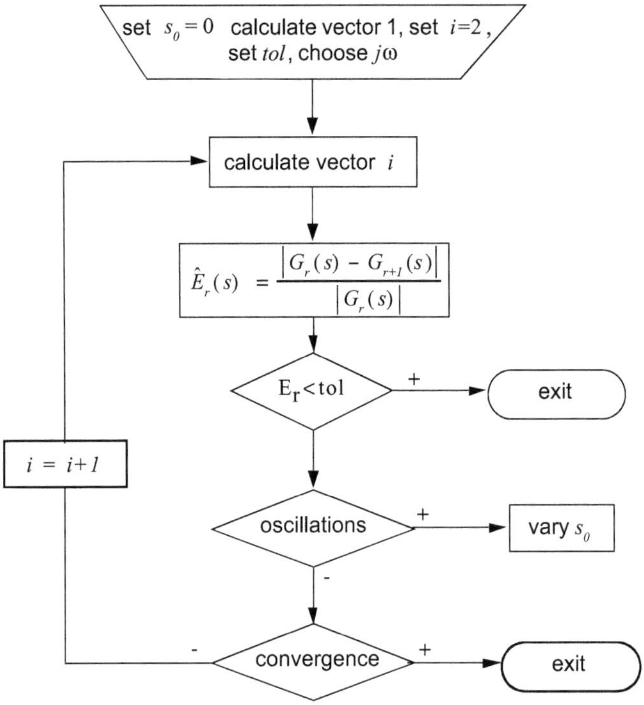

Fig. 6.15. Error indicators algorithm, based on the convergence of relative error in the frequency domain (Eq. (6.3)).

The algorithm in Figure 6.15 is similar to [168] with the advantages that it is computationally simpler and that it includes oscillations and convergence checks. Concerning the choice of the frequency range of interest, the low-pass filter characteristics of electro-thermal models can be used to connect the rise-time and the critical frequency according to Equation 5.13 for example. Another challenge is the choice of the error norm itself. The expressions in Equation 6.2 and Equation 6.5 must eventually be modified if $|G_r(s)| \approx 0$ or $y_r(t_i) \approx 0$ in order not to end up with huge relative errors. In the above examples we have made computations with temperature values in Kelvin in order to avoid such numerical problems. It would have been further possible to replace the discrete time points t_i with a continuous time variable and therefore the sum through an integral:

$$\int_{t=0}^{t_{end}} \frac{y(t) - y_r(t)}{y(t)} dt \tag{6.7}$$

but as in engineering applications the results are frequently given at a few discreet points, we have decided for simplicity to define error as in Equation 6.4.

6.1.2 Convergence of Hankel Singular Values

The second approach proposed is based on the computation of Hankel Singular Values (HSV) of the reduced system. Let us emphasize once more that the Hankel Singular Values of the linear dynamic system from Equation 3.1 of the order n can be computed by solving two Lyapunov equations, Equation 3.3 after the controllability grammian P and the observability grammian Q. The Hankel Singular Values are defined as square roots of the eigenvalues of the product of both grammians. Well-established MOR methods from Section 3.1 offer a global error bound for an approximant of order r, based on the sum of the tail of the ordered set of Hankel singular values $\Sigma = \{\sigma_1, \sigma_2, \ldots, \sigma_n\}$, $\sigma_{i+1} \geq \sigma_i$ beginning with entry $r+1$:

$$\|G - G_r\|_\infty \leq 2(\sigma_{r+1} + \ldots + \sigma_n) \tag{6.8}$$

Here $G(s)$ is the transfer function of the original state-space model Equation 5.8 and $G_r(s)$ is the transfer function of its reduced order-r model, which is obtained by using projectors originating from the solutions of Equation 3.3.

As already mentioned, unfortunately it is not possible to solve Lyapunov equations and compute HSV for large-scale systems. Luckily we have found that, for our case studies from Chapter 4, the frequency-response error of the Arnoldi reduction can be approximated by Equation 6.8 by only computing the HSV $\hat{\sigma}_{ij}$ of the reduced system in each iteration. In this way, after i iterations we have a matrix-like structure:

$$H_i = \begin{bmatrix} \hat{\sigma}_{11} & 0 & \ldots & 0 \\ \hat{\sigma}_{21} & \hat{\sigma}_{22} & \ldots & 0 \\ & \ldots & & \\ \hat{\sigma}_{i1} & \hat{\sigma}_{i2} & \ldots & \hat{\sigma}_{ii} \end{bmatrix} \tag{6.9}$$

where $\hat{\sigma}_{ij}$ is the jth HSV of the ith order reduced model. We have observed for the electro-thermal MEMS models from Chapter 4 that after a number of Arnoldi iterations the largest $\hat{\sigma}_{ij}$ of the reduced order models created converge towards the HSV of the original model. Figure 6.16 shows that for the large-scale gas sensor model the reduced system of order 50 already reproduces the original eight largest HSV. Furthermore, in each iteration one new value is added towards the end of the set, while the initial values slowly converge. This means that after a number of iterations we can consider the largest original HSV (those which do not change any more when increasing the reduced system order) as known and use Equation 6.8 to approximate the frequency response error. To demonstrate this, let us set an error bound of 10% for the transfer function of optical filter model and query the order of the reduced system needed to fulfill this error. This can be expressed as:

$$|G(s) - G_r(s)| \leq 0.1, r = ? \tag{6.10}$$

Figure 6.17 shows that after the second iteration, the first two Hankel singular Values of the optical filter seem to have converged, which means that the difference

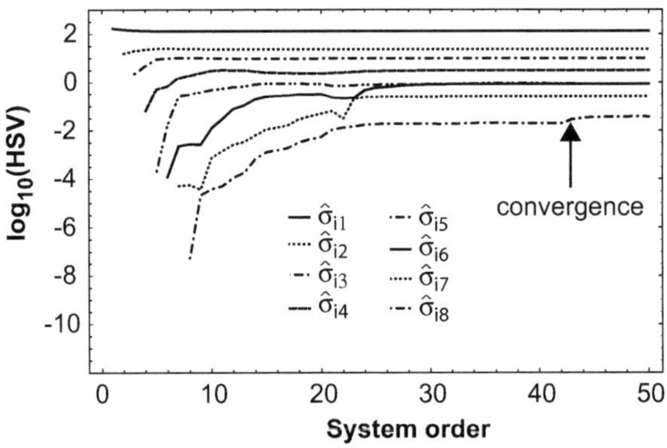

Fig. 6.16. Largest eight HSV of the Arnoldi reduced gas sensor models (order 1 to 50).

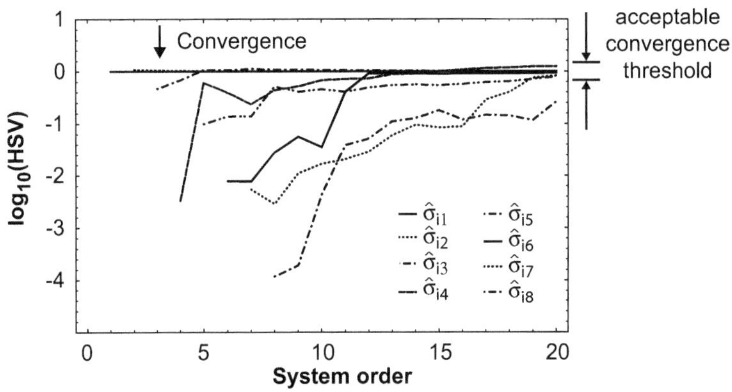

Fig. 6.17. Largest eight HSV of the Arnoldi reduced filter models (order 1 to 20) normalized to the corresponding HSV of the original model.

between two consecutive values is negligible. Hence, we can consider both as known.

Observing the matrix H_6 we further see that the third reduced HSV $\hat{\sigma}_{i3}$ converges towards $\sigma_3 = 0.04$, i.e. has an order of magnitude of 10^{-2}.

$$H_6 = \begin{bmatrix} 118.6 & 0 & 0 & 0 & 0 & 0 \\ 118 & 0.62 & 0 & 0 & 0 & 0 \\ 118 & 0.6 & 0.02 & 0 & 0 & 0 \\ 118 & 0.6 & 0.02 & 1.31\cdot 10^{-5} & 0 & 0 \\ 118 & 0.58 & 0.04 & 2.4\cdot 10^{-3} & 1.1\cdot 10^{-4} & 0 \\ 118 & 0.58 & 0.04 & 1.5\cdot 10^{-3} & 1.5\cdot 10^{-4} & 3.8\cdot 10^{-6} \end{bmatrix} \tag{6.11}$$

If we assume the worst case, that is that $\sigma_4, \sigma_5, \ldots$ are of the same order of magnitude as σ_3, and apply the estimate (6.8) we get:

$$|G(s)-G_2(s)| \leq 2 \cdot 1666 \cdot 10^{-2} \approx 33 \tag{6.12}$$

As we assume a rapid decay of the HSV for our MEMS models [6], we need some HSV decay estimate in order to correct the right side of Equation 6.12. In [175] an effective decay estimate for the eigenvalues $\lambda_i(P)$ of one grammian of the symmetric system has been proposed as

$$\frac{\lambda_k(P)}{\lambda_1(P)} \leq \left(\prod_{j=0}^{k-1} \frac{\kappa(A)^{(2j+1)/(2k)} - 1}{\kappa(A)^{(2j+1)/(2k)} + 1} \right)^2 \tag{6.13}$$

It is based only on the knowledge of the condition number κ of A, which can be computed by iterative methods [176]. Since we have observed essentially the same quality of results for the decay estimate of HSV and $\lambda_i(P)$ in our case studies, we have used a formula from Equation 6.13 to estimate an upper bound for σ_i:

$$\frac{\tilde{\sigma}_i}{\sigma_1} \leq \left(\prod_{j=0}^{i-1} \frac{\kappa(A)^{(2j+1)/(2i)} - 1}{\kappa(A)^{(2j+1)/(2i)} + 1} \right)^2 \tag{6.14}$$

and so correct the right side of the inequality from Equation 6.12. In Equation 6.14, $\tilde{\sigma}_i$ should be understood as an estimate of σ_i.

Another recently proposed estimate [177] given by

$$\frac{\lambda_k(P)}{\lambda_1(P)} = \frac{-1}{2 \cdot Re(\lambda_k)} \prod_{j=1}^{k-1} \left| \frac{\lambda_k - \lambda_j}{\lambda_k + \lambda_j} \right| \tag{6.15}$$

does not require symmetry of A, but depends on the complete knowledge of the spectrum of the system matrix, and is not practical for large-scale systems.

The estimate from Equation 6.13 as well as a decay curve of the true Hankel singular values of the tunable optical filter are shown in Figure 6.18.

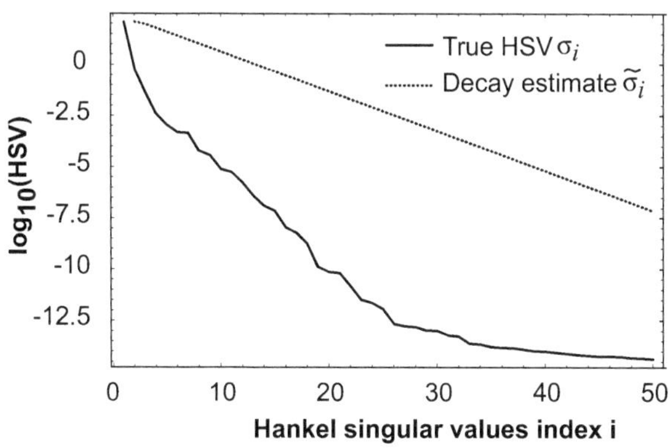

Fig. 6.18. Decay estimate of HSV for the tunable optical filter model (1668 DOF).

By adding up all the $\tilde{\sigma}_i$ estimates with an order of magnitude smaller or equal to σ_3 we get:

$$\sum_{i=j}^{1668} \tilde{\sigma}_i \approx 0.085, \tilde{\sigma}_j \le \sigma_3 \tag{6.16}$$

This sum already indicates that we could possibly fulfill our specified error bound with only two iterations. Indeed, for the original filter model it holds:

$$|G(s) - G_2(s)| \le 2 \cdot (\sigma_3 + \ldots + \sigma_{1668}) = 0.049 \tag{6.17}$$

We can repeat the whole procedure for the microthruster model. We state the task as in Equation 6.10, i.e. to search for the order of reduced model which would fulfill the 10% error. Largest 8 HSV of the Arnoldi reduced microthruster models (order 1 to 50) normalized to the corresponding HSV of the original model. Figure 6.19 shows that for the microthruster model, the reduced system of order 40 reproduces the original eight largest HSV. The matrix structure H_{40} (not displayed) shows further that the eighth reduced HSV $\hat{\sigma}_{i8}$ converges towards $\sigma_8 = 0.05$, i.e. has an order of magnitude of 10^{-2}. Assuming the worst case, that is that $\sigma_8, \sigma_9, \ldots, \sigma_{1071}$ are of the same order of magnitude as σ_8, and applying the estimate from Equation 6.8 we get:

$$|G(s) - G_7(s)| \le 2 \cdot (1071 - 7) \cdot 10^{-2} \approx 20 \tag{6.18}$$

The right side of Equation 6.18 can again be corrected through the decay estimate from Equation 6.14 by adding all the estimates $\tilde{\sigma}_i$ of σ_i which are equal or smaller than σ_8. In this way we get:

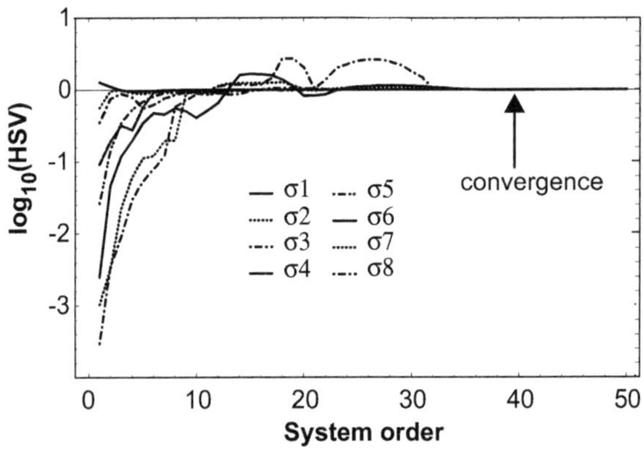

Fig. 6.19. Largest eight HSV of the Arnoldi reduced microthruster models (order 1 to 50) normalized to the corresponding HSV of the original model.

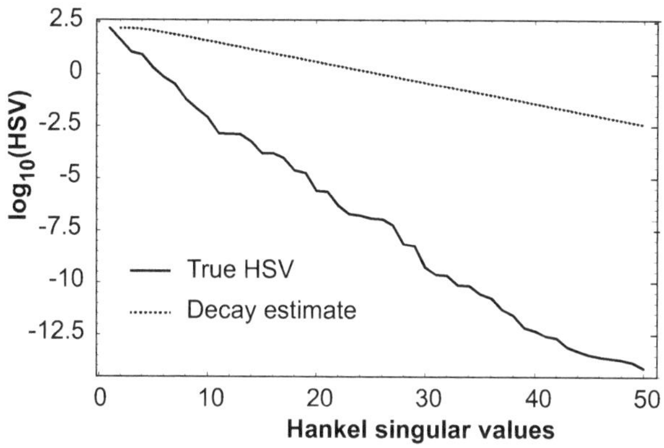

Fig. 6.20. Decay estimate of HSV for microthruster model (1071 DOF).

$$\sum_{i=j}^{1071} \tilde{\sigma}_i \approx 0.14, \tilde{\sigma}_j \leq \sigma_8 \tag{6.19}$$

The estimate from Equation 6.14 as well as a decay curve of the true Hankel singular values of the microthruster are shown in Figure 6.20.

As in the previous example, Equation 6.19 also indicates that the error bound of 0.1 could be fulfilled with seven iterations. Indeed, for the original system:

$$|G(s) - G_7(s)| \leq 2 \cdot (\sigma_8 + \ldots + \sigma_{1071}) = 0.094 \tag{6.20}$$

In the case of the large-scale model of a gas sensor it was not possible to compute true Hankel singular values, hence the estimate from Equation 6.14 was computed (Figure 6.21 shows the first 1000 estimates). Let us follow the same logic as in the 2D examples. Once again, Figure 6.16 shows that after less than 50 iterations,

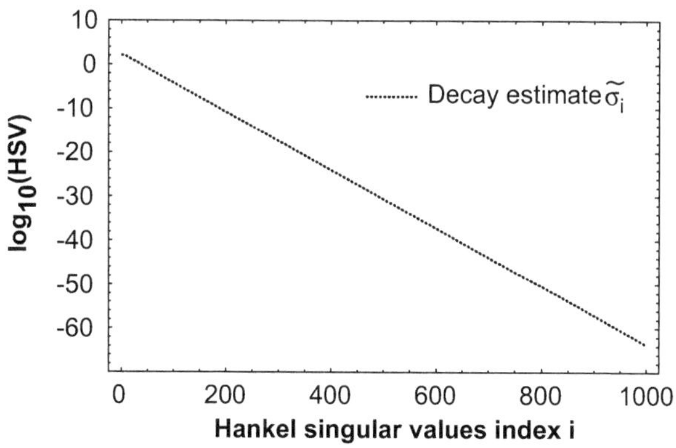

Fig. 6.21. Decay estimate of HSV for gas sensor model (73955 DOF).

the first eight Hankel singular values of the gas sensor model have converged. From the matrix H_{50} (not displayed) we find that the ninth reduced HSV $\hat{\sigma}_{i9}$ converges towards $\sigma_9 = 0.02$, i.e. has an order of magnitude of 10^{-2} . The worst case gives

$$|G(s)-G_8(s)| \leq 2 \cdot 73947 \cdot 10^{-2} \approx 147894 \tag{6.21}$$

Luckily, by adding up all the $\tilde{\sigma}_i$ estimates with an order of magnitude smaller or equal to σ_9 we get:

$$\sum_{i=j}^{73955} \tilde{\sigma}_i \approx 0.2 \, , \; \tilde{\sigma}_j \leq \sigma_9 \tag{6.22}$$

which indicates that the reduced model of order eight could have an approximate error given by:

$$|G(s)-G_8(s)| \leq 2 \cdot 0.2 \approx 0.4 \tag{6.23}$$

Let us mention that the sum in Equation 6.22 must not necessarily be made over the complete model size, because already after $i = 1000$ has been reached, the order of magnitude of $\tilde{\sigma}_i$ is only 10^{-64} . Hence, the remaining terms can be neglected.

Note once more that $G_2(s)$ in Equation 6.17, $G_7(s)$ in Equation 6.20 and $G_8(s)$ in Equation 6.23 should be computed by balanced truncations and hence the method suggested here only indicates where to stop in an iterative MOR based on the Arnoldi process. The presented strategy is schematically summarized in Figure 6.22.

The challenge in the above algorithm is to choose the right number of the largest HSV m, which must have converged if the error bound ε is to be fulfilled. In the case of the optical filter and microthruster, the convergence of two and seven largest HSV respectively was enough to fulfill an error bound of $\varepsilon = 0.1$. If one, however, needs too many iterations to meet the order of magnitude of ε among the converged HSV, the solution of the Lyapunov equations may become too time-consuming, and hence the error bound must be increased. Presently there are no guidelines on how to coordinate these numbers.

6.1.3 Sequential Model Order Reduction

The basic idea behind the sequential MOR is to first use the Arnoldi algorithm to reduce a large-scale ordinary differential equation system of order n to some order r_1, and then to switch to one of the mathematically superior control theory methods for further reduction from order r_1 to chosen order r_2. The computational effort of such an approach totals to

$$O(2r_1^2 n + 2r_1 N_z(A)) + O(r_1^3) \tag{6.24}$$

where $N_z(A)$ is a number of non-zero elements of a single system matrix A. In this way we have an exact error estimate between the reduced system of order r_1 and that

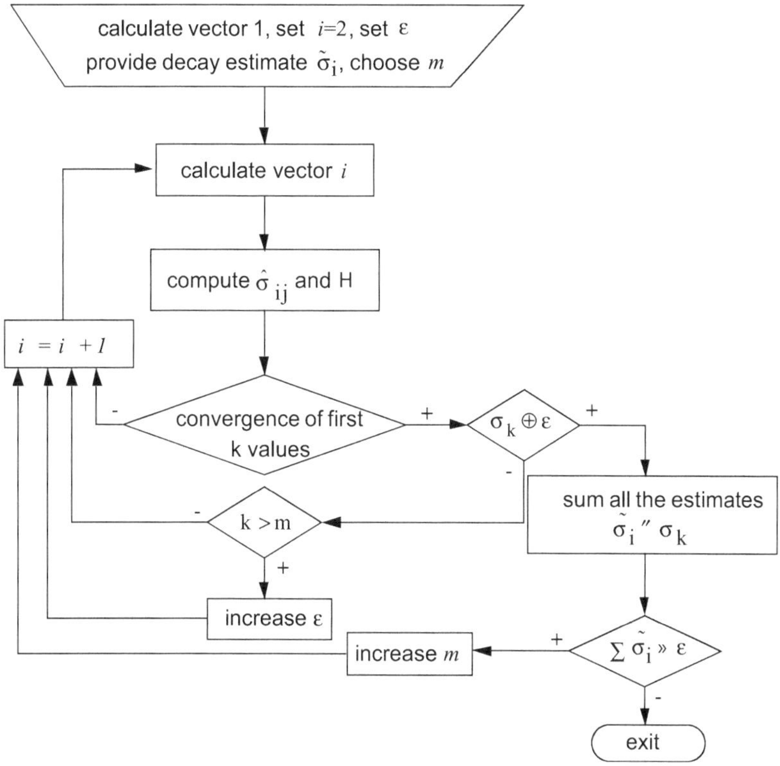

Fig. 6.22. Error indicator algorithm based on the computation of HSV of the reduced system.

of order r_2. The problem of choosing the proper r_1 remains. We suggest choosing r_1 according to one of the stop criteria from Sections 6.1.1 and Sections 6.1.2. In fact, for the tested electro-thermal MEMS models, $r_1 = 50$ turns out to be a reasonable choice, as shown in Figure 6.23, for the gas sensor model when first using Arnoldi for the reduction from 73955 to 200, 150, 100 and 50 respectively and then singular perturbation approximation. Figure 6.24 to Figure 6.26 compare the results of sequential MOR for the gas sensor model with $r_1 = 50$ and $r_2 = 5$ using the singular perturbation approximation (SPA), Hankel norm approximation (HNA) and balanced truncation approximation (BTA) to the "pure" Arnoldi reductions with $r = 50$ and $r = 5$. We observe that the target reduced order 5 can be reached with smaller error if sequential MOR is used rather than the Arnoldi algorithm alone.

To combine the error indicators from Sections 6.1.1 and 6.1.2 with sequential MOR, we should first compute the transfer function of the reduced system or the HSV in each Arnoldi iteration until some prescribed error ε_1 has been approximately fulfilled:

$$|G(s) - G_{r1}(s)| \leq \varepsilon_1 \tag{6.25}$$

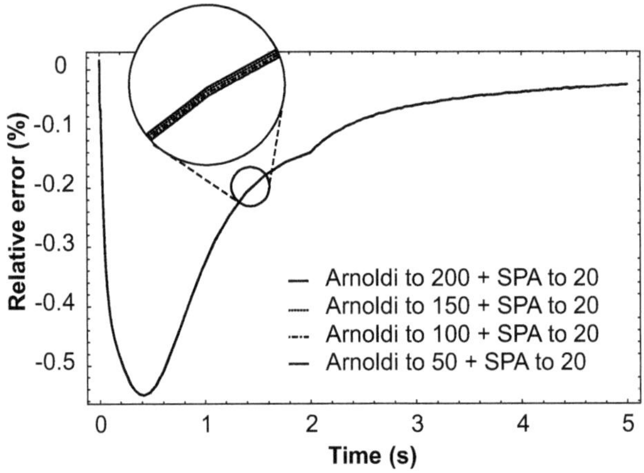

Fig. 6.23. Sequential reduction of a gas sensor using Arnoldi with r_1=200, 150, 100 and 50 and SPA with $r_2 = 20$.

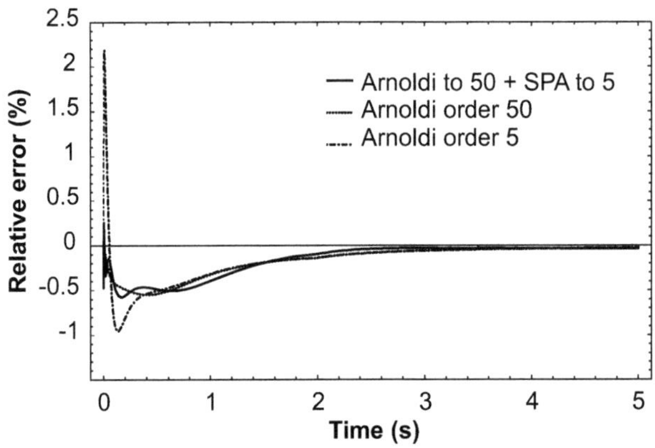

Fig. 6.24. Sequential reduction of a gas sensor using Arnoldi and SPA with $r_1 = 50$ and $r_2 = 5$.

and then use one of the three presented control theory methods to reduce the model further toward the target order r_2 having:

$$\left|G_{r1}(s) - G_{r2}(s)\right| \leq \varepsilon_2 \tag{6.26}$$

Using a simple triangle rule $\|x-y\| \geq \|x\| - \|y\|$, the error between the target reduced system with order r_2 and the full-scale system can now be expressed as:

$$\left|G(s)\right| - \left|G_{r2}(s)\right| \leq \varepsilon_1 + \varepsilon_2 \tag{6.27}$$

It should be noted that inequality from Equation 6.27 is only a guideline on how to approximately choose a reduced order model.

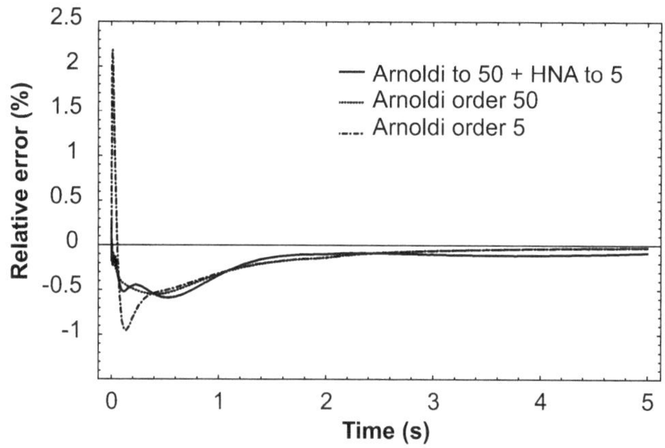

Fig. 6.25. Sequential reduction of a gas sensor using Arnoldi and HNA with $r_1 = 50$ and $r_2 = 5$.

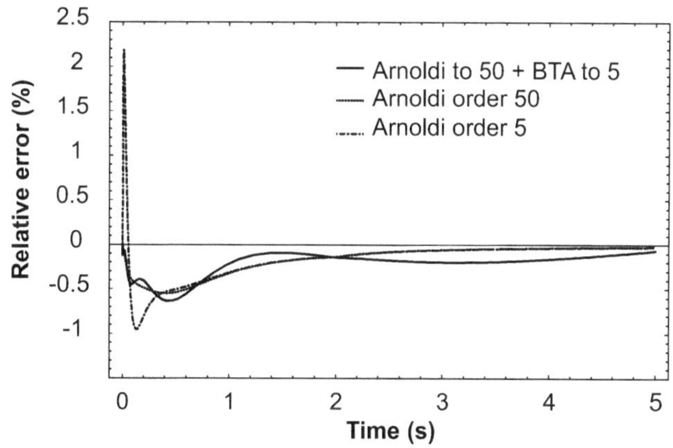

Fig. 6.26. Sequential reduction of a gas sensor using Arnoldi and BTA with $r_1 = 50$ and $r_2 = 5$.

6.1.4 Advantages and Disadvantages of the Proposed Strategies

The computation of the modeling error involves a common trade-off between computational efficiency and the accuracy of approximation. We have described three strategies and, as usual, each has its advantages and disadvantages (see Table 6.1). At the present stage, the convergence of relative error and sequential MOR can be recommended for practical use. They are both straightforward to implement.

The extra computational time required for the convergence of the relative error is very small, provided it is estimated in the frequency domain. In this case, the strategy can be summarized in two main steps:

1. Choose an error required at the highest relevant (cutoff) frequency.
2. Solve a linear equation system for this frequency at each iteration.

Table 6.1. Advantages and disadvantages of the proposed strategies.

	Convergence of relative error	Convergence of HSV	Sequential MOR
Simple implementation	+	–	+
Computational complexity	low (add. linear solve in each iteration)	high (sol. of Lyapunov eq. in each iteration)	medium (single sol. of Lyapunov eq.)
Detection whether expansion point is too far away	+	–	–
Lowest system order for the specified error	–	needs further research	+

This is the only method of the three for which it is possible to detect (through the oscillations) whether the required frequency is too far from the expansion point or not.

The computational time for sequential model reduction is higher than that for the convergence of the relative error, because first a maximum number of Arnoldi vectors (r_1) has to be estimated. Our tests show that for electro-thermal MEMS models $r_1 = 50$ might be sufficient. Now, the further reduction from r_1 to r_2 (which also has to be chosen) by BTA, SPA or HNA takes additional time. On the other hand, the order of the final reduced model might be lower than if it was determined by the convergence of the relative error for the same prescribed error bound. This is because the model of order r_2 includes information from r_1 Arnoldi vectors, although $r_1 > r_2$.

The convergence of the Hankel singular values requires the largest computational effort, because it is necessary to solve Lyapunov equations for the reduced model in each iteration. It also requires a decay estimate for the HSV. However, it is better suited to theoretical analysis than the other two methods. The advantages and disadvantages of each method are summarized in Table 6.1.

At this time we offer no theoretical justification for the proposed strategies. Whether they will function in general remains a question which requires further research, particularly with more complex transfer functions than those of the low-pass filters. We have shown, however, that they currently offer sufficient accuracy for the presented engineering problems. Due to the simplicity of implementation, the high computational efficiency and low susceptibility to numerical error accumulation, convergence of the relative error in the frequency domain followed by one of the control theory methods takes priority for the automatic generation of dynamic compact thermal models. Nevertheless, the convergence of the HSV and its application for error estimation certainly deserves further research.

6.2 Order Reduction of Interconnected Thermal MEMS Models

In Chapter 5 large-scale linear systems originating from the spatially discretized heat transfer equation, Equation 5.3 were reduced by the Arnoldi algorithm without taking into account their structure. As such systems are often composed of subsystems that are interconnected, array structures for example, it is desirable, especially with a large number of subsystems, to reduce each subsystem on its own and then to couple them back together. Hence, we seek a kind of compact thermal multiport representation which allows thermal fluxes to cross the boundaries and enables straightforward coupling to the next thermal multiport. The main problem thereby is that the thermal flow is not lumped in nature as, for example, the electrical flow is along metallic wire interconnects. The ratio of electrical conductivity of metals and that of insulators is of the order of 10^8. Hence, electrical current flow takes place almost solely in metal paths. This is not the case with heat flow because the ratio of thermal conductivities in microtechnology is only of the order of 10^2 (see Figure 6.27). Therefore, it is unclear how to lump the thermal fluxes at shared surfaces between two finite element (FE) models in order to form the thermal ports (Figure 6.28) which serve to couple together several compact models. As a matter of fact, there appears to be very few studies [178, 179] into how to couple (dynamic) compact thermal models.

Below, we present and discuss the currently available solutions for MOR of thermal micro-array structures. In Sections 6.2.1 the MEMS case study, a MOS-transistor-based microhotplate array model, is presented. It is a high-dimensional model, containing several hundred FE nodes at each shared interface. Section 6.2.2 describes the possibility of reducing the entire array, i.e. without decoupling the parts, via the block Arnoldi procedure, which is suitable for multiple-input–multiple-output (MIMO) linear systems. It offers a simple and effective solution for the reduction of array structures, but "hides" the problem of coupling the dynamic compact thermal models. Section 6.2.3 discusses the possibility of coupling reduced order models via substructuring based on a modified Guyan algorithm. It is usually used in structural mechanics, but is available for the thermal domain as well. As it demands the preservation of all shared (coupling) nodes, the resulting orders of compact models are much larger than by block Arnoldi. Section 6.2.4 addresses some possibilities for coupling reduced models in the general case via additional flux inputs and introduces a structure-preserving MOR. In Sections 6.2.5 we give a brief comparison of the available methods.

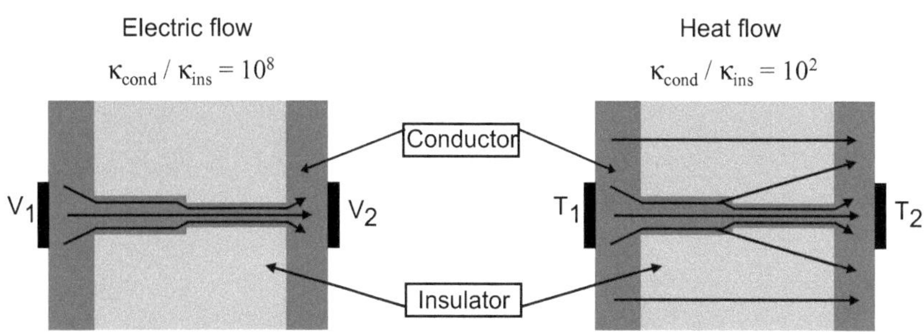

Fig. 6.27. Comparison of distributive effects: electric flow is lumped and heat flow is not.

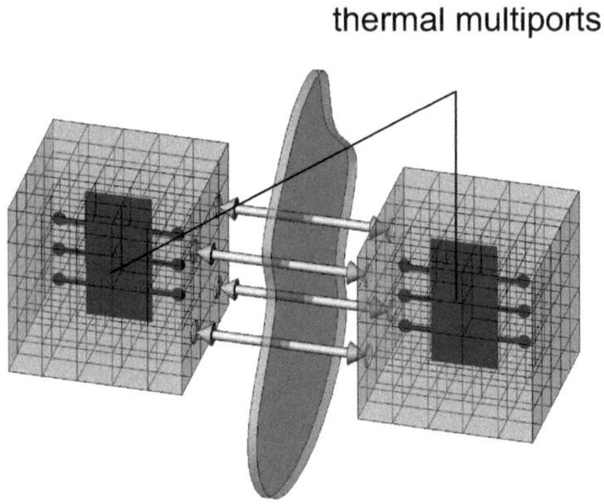

Fig. 6.28. Continuous thermal flux through the shared interface of two FE models. The goal is to model the "FE cubes" as thermal multiports, i.e. to "lump" the flux.

6.2.1 Microhotplate Array

A microhotplate array model is built upon a real gas sensor device fabricated in industrial CMOS technology [26, 36]. It is composed of a $500\,\mu m \times 500\,\mu m$ CMOS multilayer fully suspended membrane and a ring PMOS transistor heater of $5\,\mu m$ gate length and $720\,\mu m$ overall gate width buried in a silicon island under the membrane, as shown in Figure 6.29. Figure 6.30 shows an example of a fabricated chip with digital circuitry in the lower part, analog circuitry in the middle and an array of three microhotplates in the upper section. The microhotplates are covered with drop-deposited noncrystalline SnO_2 as sensitive layers. Full advantage is taken of the features offered by applying CMOS technology. All sensor values can be set and read out via the digital interface, which drastically reduces the packaging complexity since the number of bond wires is the same as for a single microhotplate.

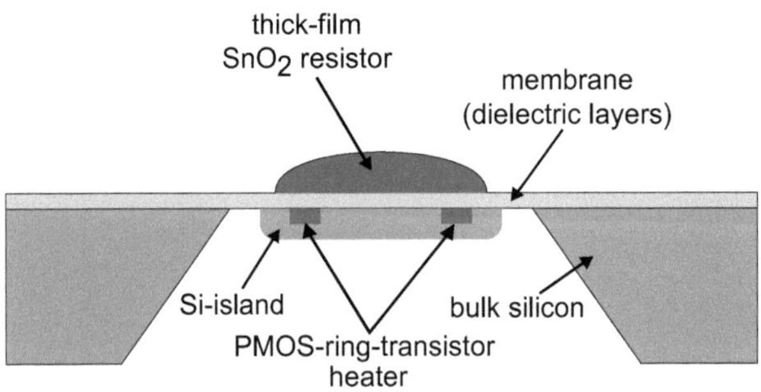

Fig. 6.29. Cross-section of a single microhotplate.

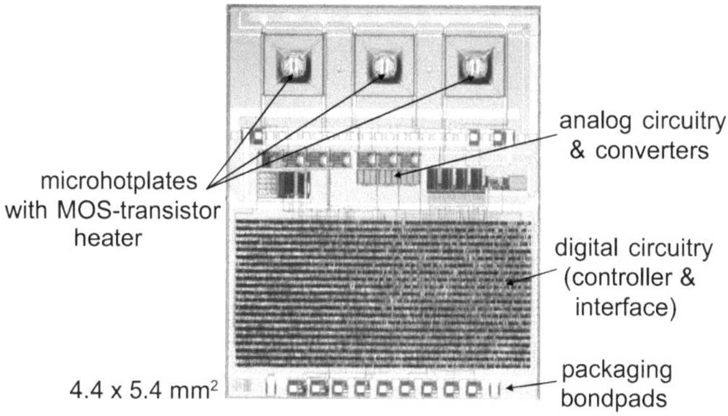

Fig. 6.30. Micrograph of the chip with microhotplate array and circuitry. Picture courtesy of M. Graf (ETH Zürich, Switzerland).

Before the microhotplate devices are fabricated, the designs undergo an extensive simulation process, such as thermal modeling, using finite element simulations. Modeling the transducer and hotplate behavior is an extremely important step towards the formulation of a compact sensor model for monolithic system realizations. The parameters of interest include the hotplate thermal resistance and thermal time constant, prediction of which facilitates and accelerates the design of the monolithic array systems for a given microhotplate design.

For test purposes, we built a finite element model of 2×2 microhotplate array structure which contains 100 934 DOF (meshed and discretized with ANSYS using SOLID90 elements) and several hundred FE nodes at each shared surface. We modeled the membrane as a single layer with material properties as proposed in [180]. The transistor heater is modeled as a lumped element circular polysilicon heater buried in a Si island under the membrane, which allows for a pure thermal simulation. Our model captures the thermal efficiency of the real device, which is $6°C/mW$ [181], to within 70%.

For the reduction of interconnected thermal systems it is necessary to describe the thermal crosstalk within a reduced order model. To accomplish this, we modelled a 2×2 array as a part of a larger array. At all four side walls of the chip, the convection boundary conditions, as defined in Equation 2.9 were applied. The heat transfer coefficient was set to $h = 10^4$ and the ambient temperature was set to $T_{ambient} = 0°C$. The initial temperature was also set to zero. The heat source power for each circular heater was set to $40\,mW$. In Section 6.2.2 it will be shown that it is possible to turn heat sources on and off also after the MOR step. Temperature contour plots in Figure 6.31 through 6.33 demonstrate that the chosen BCs result in a temperature increase of about $10°C$ per added heat source everywhere on the chip. This effect must be taken into account if no automatic temperature control is employed. Note, however, that the model only resembles the real device and hence the true crosstalk level may be different.

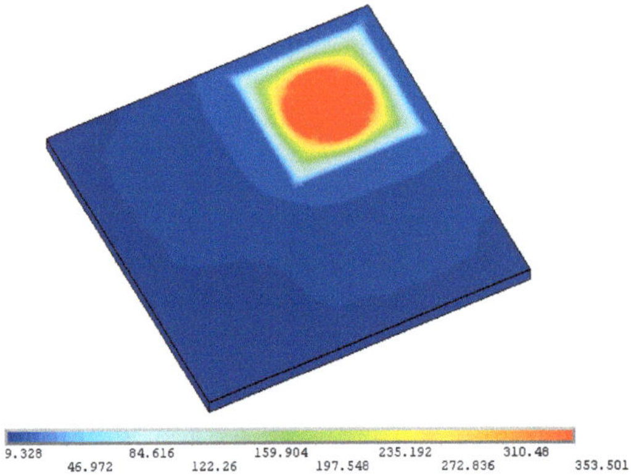

Fig. 6.31. Steady-state temperature distribution in °C of a 2×2 microhotplate array with a single heat source.

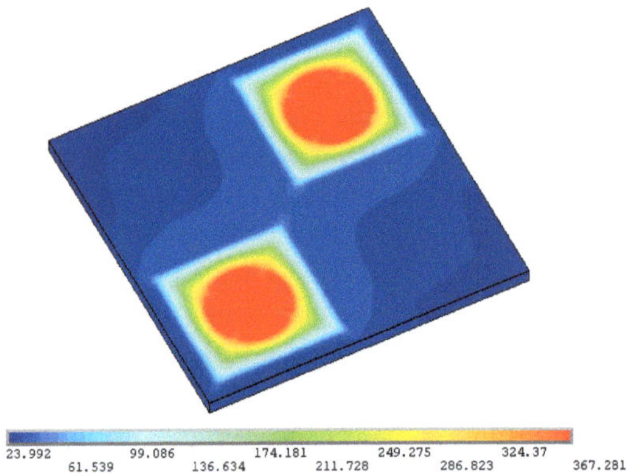

Fig. 6.32. Steady-state temperature distribution in °C of a 2×2 microhotplate array with two heat sources.

6.2.2 Block Arnoldi

A block Arnoldi method [103] offers a straightforward approach to reduce a model of an array structure with multiple heat sources, provided that the number of devices within an array remains moderate (see the discussion at the end of the section). With block Arnoldi the entire array structure is meshed and reduced in a similar way to a classical single-input–single-output Arnoldi process. Let us briefly explain this.

The multiple-input–multiple-output heat transfer equation in discretized form is given by:

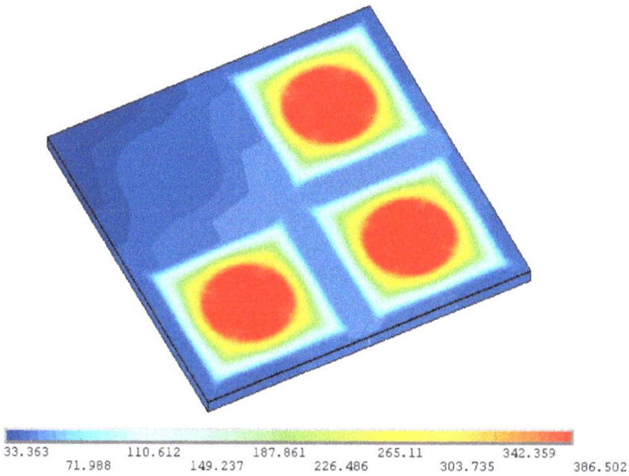

Fig. 6.33. Steady-state temperature distribution in °C of a 2×2 microhotplate array with three heat sources.

$$\begin{aligned} C\cdot\dot{T}+K\cdot T&=F\cdot Q(t) \\ y(t)&=E^T\cdot T \end{aligned} \tag{6.28}$$

where $F\in R^{n\times m}$ and $E\in R^{n\times p}$ are the input and the output matrix, and m and p denote the number of inputs and outputs, respectively. Unlike in Equation 5.3 the inputs $Q(t)=[Q_1\,Q_2\ldots Q_m]^T\to R^m$ and the outputs $y[0,\infty)\to R^p$ are both vector-valued functions. As a result, a transfer function of Equation 6.28 is a matrix-valued rational function $G{:}C\to C^{p\times m}$, which is analog to Equation 5.8, given by

$$G(s)=E^T\cdot(K+sC)^{-1}\cdot F, s\in C \tag{6.29}$$

The block Arnoldi method obtains the information of the leading Taylor coefficients of $G(s)$, i.e. moments around a chosen frequency s_0 (zero in our case), in a similar manner to the standard SISO case. Before the block Arnoldi can be employed, the two matrices C and K have to be reduced to a single matrix, denoted by A in the following. This can be done by rewriting Equation 6.29 as follows:

$$G(s)=-E^T\cdot(I-s\cdot A)^{-1}\cdot B \tag{6.30}$$

where $A=-K^{-1}C$ and $B=-K^{-1}F$. m columns of the matrix $B=[b_1\,b_2\ldots b_m]$ are the starting vectors of the so-called block Krylov subspace, which will be defined below. We start by defining the right block Krylov matrix as

$$K^R(A,B)=\left[B\;AB\;A^2B\;\ldots\;A^{n-1}B\right] \tag{6.31}$$

Note that the left block Krylov matrix, which is needed in two-sided approaches, would have been induced by A^T and E, analogous to Equation 3.26.

Our goal is to define a sequences of ascending r-dimensional subspaces, $r = 1,2,\ldots$, that are spanned by the first r linearly independent columns of the matrix $K^R(A,B)$. To properly define these subspaces, we need to delete the linearly dependent columns in the right-hand side of Equation 6.31. For example, if a column $A^j b_i$ in Equation 6.31 is linearly dependent on earlier columns, than this column and all its successive A-multipliers need to be deleted. Consequently, we obtain a so-called deflated block Krylov matrix:

$$K^{defl}(A,B) = \left[B_1\, AB_2\, A^2B_3 \ldots B_{j_{max}}\right] \tag{6.32}$$

where for each $j = 1,2,\ldots,j_{max}$, B_j is a submatrix of B_{j-1} if deflation (the process of deleting linearly dependent vectors) occurred within the jth Krylov block $A^{j-1}B$. Here, for $j = 1$ we set $B_1 = B$. Hence, if we denote by m_j the number of columns of B_j, we have:

$$m = m_1 \ge m_2 \ge \ldots \ge m_{j_{max}} \ge 1 \tag{6.33}$$

The rth right block Krylov subspace induced by A and B, $K^R(A,B)$, is the subspace spanned by the first r columns of the right-hand side in Equation 6.32:

$$K_r^R(A,B) = colspan\{B_1, AB_2, \ldots, A^{j-1}B_j\} \tag{6.34}$$

Therefore, $r = m_1 + m_2 + \ldots + m_j$ with $1 \le j \le j_{max}$. We remark that, for a single starting vector $B = b$ block Krylov subspace, Equation 6.34 is identical to Equation 5.6. The block Arnoldi process (Algorithm 7.1 [103]) extends the classical Arnoldi algorithm to block Krylov subspaces $K_r^R(A,B)$, $r \ge 1$. Some variations of the block-Arnoldi algorithm can be found in [182].

The reduction of the original system from Equation 6.28 is performed by projection, as in Equation 5.7:

$$\begin{aligned} V^TCV\cdot\dot{z} + V^TKV\cdot z &= V^TF\cdot Q(t) \\ y_r(t) &= E^TV\cdot z \end{aligned} \tag{6.35}$$

Let us go back to our case study. For the 2×2 microhotplate array with four heat sources and four defined outputs, the multiple input vector $Q(t) = [Q_1\ Q_2\ Q_3\ Q_4]^T$ does not takepart in Algorithm 6.1 and hence it is possible to switch each heat source on and off after the order reduction. Figure 6.34 and Figure 6.35 compare the step responses of the full-scale and reduced order 50 model, computed by block Arnoldi, for the case when two heat sources (each 40 mW) are switched on. The observed outputs are located in the center of each hotplate. Figure 6.36 shows further that the reduced model accurately reproduces the crosstalk effect presented in the previous section.

Algorithm 6.1

Inputs: A and B

Outputs: *projection matrix* V *and the matrix of the reduced system* $H = V^T A V$

0. Start:
Set $\hat{v}_k = b_k$, *for* $k = 1, 2, \ldots, m$.
Set $m_c = m$ *and choose a deflation threshold* $dtol \approx 0$.

For $i = 1, 2, \ldots, r$ *do:*
1. Deflation:
If $\|\hat{v}_i\| \le dtol$ *set* $m_c = m_c - 1$.
If $m_c = 0$ *set* $i = i - 1$ *and stop.*
If $m_c \neq 0$ *set* $\hat{v}_k = \hat{v}_{k+1}$ *for* $k = i, i+1, \ldots, i+m_c-1$.
Return to step 1.

2. Normalize:
$v_i = \hat{v}_i / \|\hat{v}_i\|$

3. Compute the next vector from the new block:
$\hat{v}_{i+m_c} = A \cdot v_i$

4. Orthogonalize:
For k=1,2,,,i
set $h_{k,i} = v_k^T \hat{v}_{i+m_c}$ *and* $\hat{v}_{i+m_c} = \hat{v}_{i+m_c} - v_k h_{k,i}$
For $k = i - m_c + 1, i - m_c + 2, \ldots n - 1$
set $h_{i,k} = v_i^T \hat{v}_{k+m_c}$ *and* $\hat{v}_{k+m_c} = \hat{v}_{k+m_c} - v_i h_{i,k}$.

End for i loop.

One may ask the question if the block Arnoldi could have been avoided by using four times the standard Arnoldi with Krylov subspaces induced by single vectors

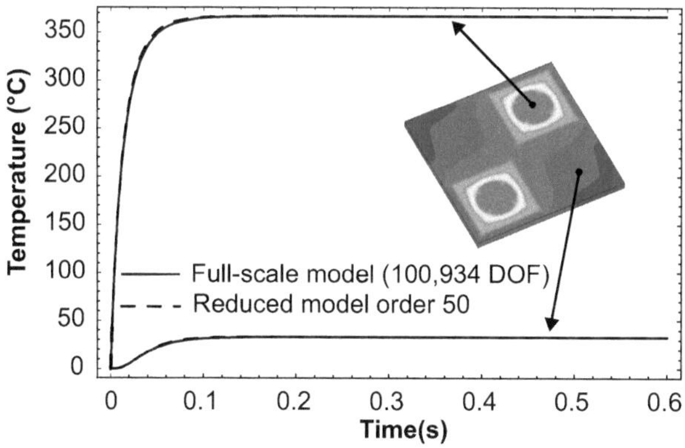

Fig. 6.34. Step response of the full-scale and reduced order models at two output points when two heat sources of 40 mW each are switched on.

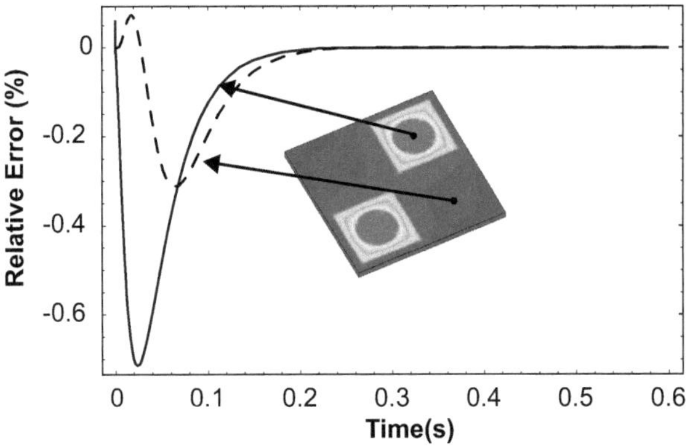

Fig. 6.35. Step response errors corresponding to Fig. 7.8.

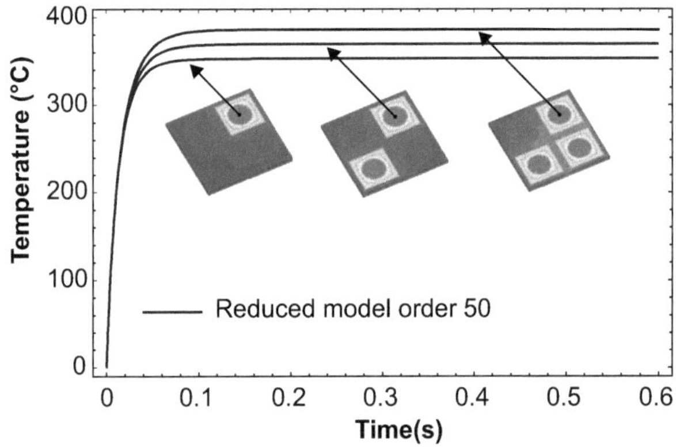

Fig. 6.36. Step response of the reduced order 50 models at a single output point when one, two or three heat sources of 40mW each are switched on.

instead. For example, one could generate scalar approximations for all sixteen coefficient functions of the 4×4 matrix valued transfer function from Equation 6.29 via a suitable basis for standard Krylov subspaces $K_r(A,b_1)$, $K_r(A,b_2)$, $K_r(A,b_3)$ and $K_r(A,b_4)$. In this way instead of a single reduced model from Equation 6.35 with four heat sources and output temperature vector $y_r(t)$, we would have four reduced models, each describing a complete output (speciality of Arnoldi process) for a single heat source. Their outputs can be denoted $y_{r1}(t),\ldots,y_{r4}(t)$. In the case when, for example, the first and fourth heat sources are on, the chosen output temperatures from Equation 6.35 can be computed as a superposition of SISO models outputs:

$$y_r(t) = y_{r1}(t) + y_{r4}(t) \tag{6.36}$$

in each single node. The standard Lanczos algorithm, on the other hand, would have to be used sixteen times due to the approximation of a single output only. However, in [103] it is stated that block Krylov methods for MIMO systems result in more efficient MOR than those based on standard Krylov subspaces. This issue certainly deserves more research.

It seems so far that block Arnoldi offers a simple and effective solution for the MOR of interconnected MIMO systems. Its main advantage over a substructuring method (described in Sections 6.2.3) is that no master degrees of freedom have to be chosen and the minimum size of the reduced model is not influenced by the large number of interface nodes. As the output matrix E does not takepart in Algorithm 6.1, the approximation of the complete output ($E = I_{n\times n}$) is warranted as in the SISO case (see Sections 5.3.1). A further advantage is that the reduced system from Equation 6.35 can be very easily formally transferred into a HDL model, as will be demonstrated in the following section. The disadvantage of block Arnoldi becomes evident with the growing number of devices within an array. At some point, the system matrices C and K grow so large that the LU or Cholesky decomposition needed for construction of the projection matrix may become difficult or even prohibitive. In such cases, it is clearly better to search for an alternative to reduce each device individually (using either projection or modal analysis) and then to couple them. How this can be done formally in the general case, as well as in some special cases, will be discussed in the following sections.

Interconnected System Behavioral Simulation

We present here the HDL form of the system from Equation 6.35 and demonstrate that the crosstalk effect is described sufficiently well. In the system-level model it becomes obvious that control of the heat sources is independent of the reduced model. Block Arnoldi projection is schematically shown in Figure 6.37. A reduced system is defined with $K_r = V^T \cdot K \cdot V$, $C_r = V^T \cdot C \cdot V$ and $F_r = V^T \cdot F$. Figure 6.38 shows a HDL model for the 2 by 2 microhotplate model reduced by block Arnoldi. A template containing a reduced ODE system from Equation 6.35 is back coupled to each one of the four heaters through their input power $Q_1, \ldots, Q_4$. The observed output temperatures $T_{\text{out},1}, \ldots, T_{\text{out},4}$, at the center of four hotplates and the resistivity control temperatures $T_1, \ldots, T_4$ are computed as linear combinations of the reduced states:

$$\begin{aligned} T_i &= V^* \cdot z \\ T_{\text{out},i} &= E^T V \cdot z \end{aligned} \tag{6.37}$$

The reduced model contains no nodes in the physical sense and hence the template region in Figure 6.38 is to be understood as a visualization of the equation system from Equation 6.35.

Figure 6.39 compares the numerical simulation results of the behavioral MAST model with those of the full-scale model integrated in ANSYS and the reduced order model from Equation 6.35 integrated in Mathematica. Again, the difference between the Mathematica and SABER solutions totals to the numerical integration errors between the different solvers.

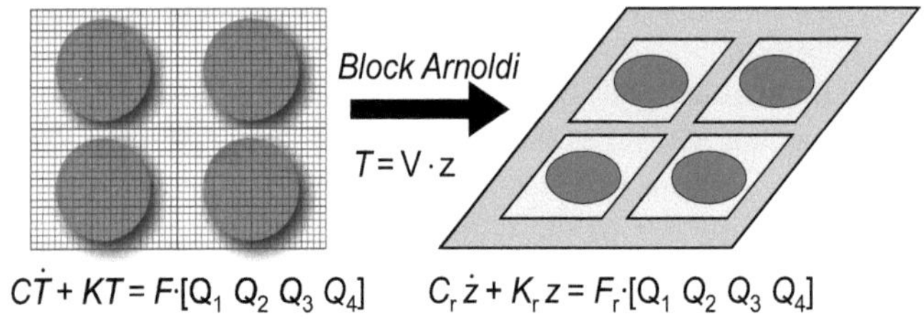

Fig. 6.37. Schematic of the Block Arnoldi projection for the microhotplate array model.

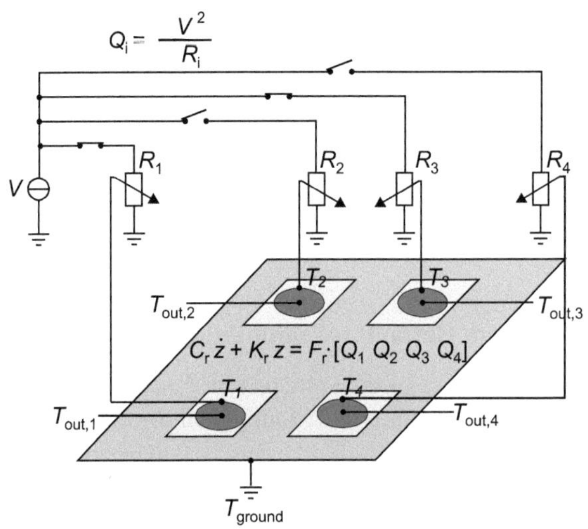

Fig. 6.38. Structure of the implemented HDL model of the microhotplate array.

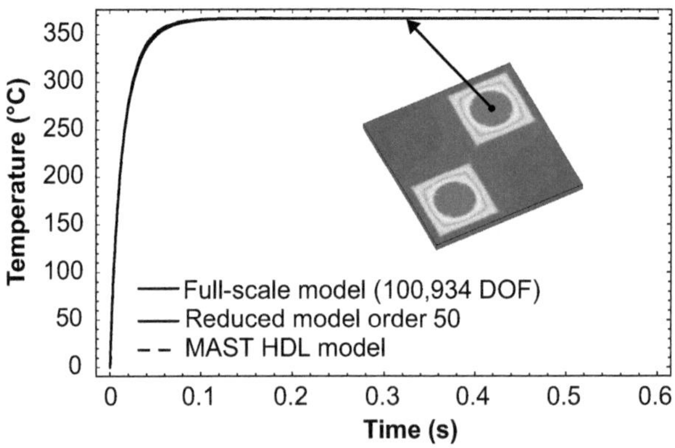

Fig. 6.39. Step response of the full-scale (100 934 DOF) and reduced order 50 model at a single output node. The reduced system was integrated in Mathematica 4.1 and SABER (MAST HDL model) and the full-scale system with ANSYS7.1.

6.2.3 Coupling of Reduced Order Models via Substructuring

This commercially available method [112] is based on modification of the Guyan algorithm, which was described in Section 3.3. The disadvantages of this algorithm when applied to order reduction of electro-thermal models have been demonstrated in Sections 5.4. Its major advantage, however, is that it allows the coupling of several reduced models into an array structure, due to the physical preservation of the shared surface nodes during reduction. The equation system which describes two thermal models, RM_1 of order r_1 and RM_2 of order r_2, reduced by a modified Guyan algorithm using Equation 3.50 and Equation 3.58, is given as:

$$\begin{aligned} C_{r1} \cdot \dot{T}_{e1} + K_{r1} \cdot T_{e1} &= F_{r1} \cdot Q_1(t) \\ C_{r1} \cdot \dot{T}_{e2} + K_{r2} \cdot T_{e2} &= F_{r2} \cdot Q_2(t) \end{aligned} \tag{6.38}$$

where $T_{e1} \in R^{r_1 \times 1}$ and $T_{e2} \in R^{r_2 \times 1}$ are terminal node vectors of RM_1 and RM_2 (see Figure 6.40). By performing a Guyan reduction of each submodel, the chosen terminal nodes T_{e1} and T_{e2} must contain at least those surface nodes which are to be coupled to another model. Depending on the desired accuracy and efficiency, additional nodes can be chosen as master degrees of freedom. In Figure 6.40 the shared nodes are denoted $T_{e1,i}$ and $T_{e2,i}$.

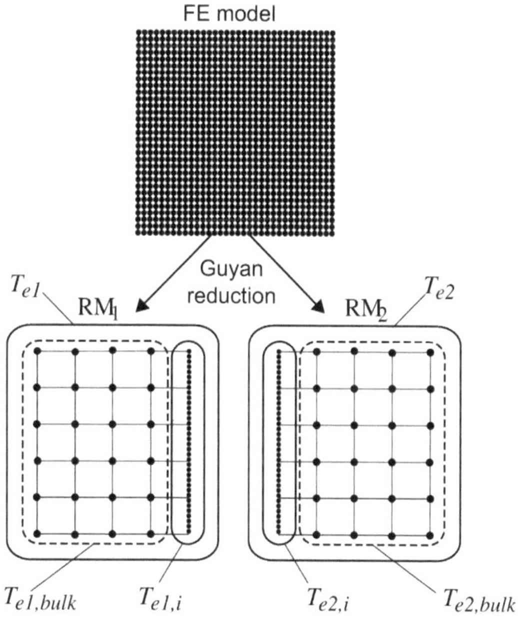

Fig. 6.40. Two Guyan-based reduced models with terminal nodes T_{e1} and T_{e2} which can be coupled over the surface nodes $T_{e1,i}$ and $T_{e2,i}$.

The coupling is done through the following set of $2\,m$ constraint equations for the temperatures and fluxes on the shared interface:

$$\begin{aligned} T_{e1,i} - T_{e2,i} &= 0, 1 \le i \le m \\ \dot{T}_{e1,i} - \dot{T}_{e2,i} &= 0, 1 \le i \le m \end{aligned} \tag{6.39}$$

Equation 6.39 reduces the number of variables in Equation 6.38 by m. The same coupling procedure can be used for coupling two full-scale models or one reduced and one full-scale model. We now transform the equation system from Equation 6.38 into a block-matrix form:

$$\begin{bmatrix} C_{11} & C_{12} \\ C_{21} & C_{22} \end{bmatrix} \cdot \begin{bmatrix} \dot{T}_1 \\ \dot{T}_2 \end{bmatrix} + \begin{bmatrix} K_{11} & K_{12} \\ K_{21} & K_{22} \end{bmatrix} \cdot \begin{bmatrix} T_1 \\ T_2 \end{bmatrix} = \begin{bmatrix} F_1 \\ F_2 \end{bmatrix} \tag{6.40}$$

where

$$T_1 = T_{e1,i} \tag{6.41}$$

are the coupling nodes belonging to RM_1 and T_2 are all other nodes belonging to both models:

$$T_2 = T_{e1,bulk} \cup T_{e2,bulk} \cup T_{e2,i} \tag{6.42}$$

This assignment is arbitrary, i.e. it is only important that m different nodes, out of $2m$ shared nodes, are gathered within a single vector (T_1 in this case). For simplicity the input terms $Q_1(t)$ and $Q_2(t)$ have been incorporated in the load vector. Let us further rewrite the constraint equations Equation 6.39 as $T_1 = X \cdot T_2$ and $\dot{T}_1 = X \cdot \dot{T}_2$ or:

$$\begin{aligned} \begin{bmatrix} I & -X \end{bmatrix} \cdot T &= 0 \\ \begin{bmatrix} I & -X \end{bmatrix} \cdot \dot{T} &= 0 \end{aligned} \tag{6.43}$$

where I is the unity matrix of dimension $m \times m$ and $T = \begin{bmatrix} T_1 & T_2 \end{bmatrix}^T$. In a general case the right-hand side of Equation 6.43 may be different from zero.

The problem of solving Equation 6.40 under the constraints Equation 6.43 can be carried out by several methods [183, 184]. In ANSYS substructuring, the Lagrange multipliers adjunction method [185] is used with the goal of minimizing the energy function. The total energy of the thermal model is given as

$$E = \frac{1}{2}\dot{T}^T \cdot C \cdot \dot{T} + \frac{1}{2}T^T \cdot K \cdot T - T^T \cdot F \tag{6.44}$$

where C and K are the system matrices and F is a load vector from Equation 6.40. To impose the constraints from Equation 6.43, we adjoin $2m$ Lagrange multipliers collected in vectors λ_1 and λ_2 and form a Lagrangian:

$$L(T,\dot{T},\lambda_1,\lambda_2)=E+\lambda_1^T\cdot[I-X]\cdot\dot{T}+\lambda_2^T\cdot[I-X]\cdot T \tag{6.45}$$

The minimization of $L(T,\lambda)$ with respect to T, $\dot{T}$, λ_1 and λ_2 yields the multiplier-augmented form of Equation 6.40:

$$\begin{bmatrix} C_{11} & C_{12} & I & 0 \\ C_{21} & C_{22} & -X^T & 0 \\ I & -X & 0 & 0 \\ 0 & 0 & 0 & 0 \end{bmatrix}\cdot\begin{bmatrix} \dot{T}_1 \\ \dot{T}_2 \\ \dot{\lambda}_1 \\ \dot{\lambda}_2 \end{bmatrix}+\begin{bmatrix} K_{11} & K_{12} & 0 & I \\ K_{21} & K_{22} & 0 & -X^T \\ 0 & 0 & 0 & 0 \\ I & -X & 0 & 0 \end{bmatrix}\cdot\begin{bmatrix} T_1 \\ T_2 \\ \lambda_1 \\ \lambda_2 \end{bmatrix}=\begin{bmatrix} F_1 \\ F_2 \\ 0 \\ 0 \end{bmatrix} \tag{6.46}$$

By eliminating T_1 and $\lambda_1+\lambda_2$ from Equation 6.46 the new equation system with dimension r_1+r_2-m, which describes two reduced coupled devices, is given through

$$\tilde{C}\cdot\dot{T}_2+\tilde{K}\cdot T_2=\tilde{F} \tag{6.47}$$

where

$$\begin{aligned} \tilde{C}&=X^T\cdot C_{11}\cdot X+X^T\cdot C_{12}+K_{21}\cdot X+C_{22} \\ \tilde{K}&=X^T\cdot K_{11}\cdot X+X^T\cdot K_{12}+K_{21}\cdot X+K_{22} \\ \tilde{F}&=X^T\cdot F_1+F_2 \end{aligned} \tag{6.48}$$

It is worth noting that at this stage it is still possible to turn the heat sources on and off, i.e. to scale the load vectors by scaling F_{r1} and F_{r2} in Equation 6.38 and hence, F_1 and F_2.

After solving Equation 6.47 it is possible to recover T_1 through Equation 6.43 and gradually via Equation 6.38 to split the nodes to the original reduced models, thus expanding finally the complete temperature field of a single device via Equation 3.48.

Let us go back to our case study. In order to accurately describe the heating area of the single hotplate during the reduction, we have preserved the 948 heater nodes additionally to the surface nodes. As a result, a reduced model of a single microhotplate, which is described by Equation 3.59, contains 2140 ordinary differential equations. The lower limit of the reduced model when applying this method would have been the description of the interfaces, which is 1192 ODEs. The coupling of four reduced models into an array structure according to Equation 6.47 and Equation 6.48 results in an equation system of 7351 ODEs. Note once more that the full-scale FE model contains 100 934 equations. An important computational issue here is that the reduced system from Equation 6.47 is dense and hence its computation may not bring along a large decrease in CPU time versus computing the full-scale system, which is sparse.

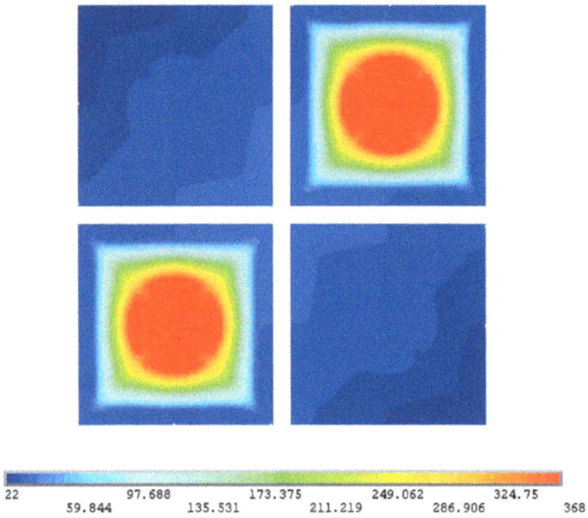

Fig. 6.41. Expanded solution of four coupled reduced models when two heat sources of 40 mW each are switched on.

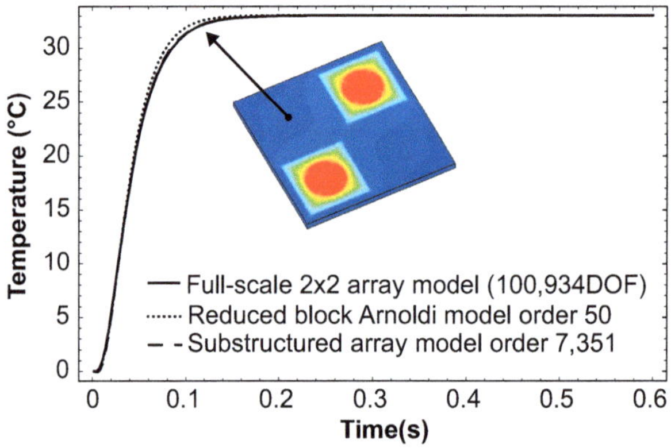

Fig. 6.42. Single output step responses of the full-scale and reduced order models created by block Arnoldi and Guyan-based substructuring. Two heat sources of 40mW are switched on.

Figure 6.41 shows the assembled contour plots of four substructured and coupled microhotplates when two heat sources are switched on. The approximation of the thermal crosstalk effect, which takes place within a full-scale FE model (Fig. 6.32.), is excellent. Figure 6.42 compares further the step responses of the substructured array with the step response of the full-scale model and the step response of the block Arnoldi reduced model from Figure 6.34.

The convenience of the Guyan method is that it preserves the surface nodes in the physical sense, while using the projection matrix as defined in Equation 3.52. Hence, it is possible to decouple an array, to reduce each device on its own, and to couple several reduced models by using any of the methods for treating linear multifreedom constraints in finite element systems [184]. The major disadvantage, however, is that the lower limit of the single reduced model is the description of the interfaces, which implies, with accurate meshing, relatively large model sizes. For the case studies presented in this thesis, several hundred nodes per single surface were accounted for. It is of course possible to reduce the number of coupling nodes by coarsening the mesh on the interface, but this results in a more difficult mesh generation and lower mesh quality. Besides, we may need to choose more bulk (nonterminal) nodes in order to reach the desired accuracy (see the results in Sections 5.4). Hence, the best approach is to use the Krylov-subspace-based reduction of a single device because it guarantees smaller dimensions and higher precision than Guyan-based substructuring and then to couple several reduced models into an array structure. Unfortunately, this is possible only in some special cases or under special approximations, as will be shown in the next section.

6.2.4 Coupling of Reduced Order Models in the General Case

In Sections 6.2.2 and 6.2.3 two state-of-the-art methods for MOR of interconnected thermal models were applied to the microhotplate array model. As usual, both have advantages and disadvantages. The main difference between them is that block Arnoldi reduces the entire system model, and substructuring allows the model to be decoupled and reduces each subsystem. Hence, block Arnoldi may not be computationally effective in cases when the number of subsystems is very large. Because substructuring preserves all shared nodes within the reduced system, it may result in unnecessarily large reduced models. In this section we discuss the possibilities of generalizing both approaches. The first possibility is to reduce each submodel without preserving the shared nodes, i.e. with Arnoldi, and to try to couple the reduced models back into a reduced array model. We will show that under some circumstances this is possible. The second possibility is to reduce each subsystem with Arnoldi without even decoupling the array model, which is of course, the best solution. We will show that the newest mathematical theories are developing in this direction.

Coupling by Fluxes

If MOR of a single device model is done by projection, as is the case in Krylov subspace methods, the surface nodes are not preserved any more. Instead, we have generalized coordinates z. In this case the Lagrange multipliers method can not be applied and therefore the only way to couple the reduced models back into an array structure is over additional "flux inputs". Let us explain this.

The two, still uncoupled models (M_1 and M_2 in Figure 6.43) are described through the equation system:

$$\begin{bmatrix} C_{11} & 0 \\ 0 & C_{22} \end{bmatrix} \cdot \begin{bmatrix} \dot{T}_1 \\ \dot{T}_2 \end{bmatrix} + \begin{bmatrix} K_{11} & 0 \\ 0 & K_{22} \end{bmatrix} \cdot \begin{bmatrix} T_1 \\ T_2 \end{bmatrix} = \begin{bmatrix} F_1 \cdot u_1 \\ F_2 \cdot u_2 \end{bmatrix} \tag{6.49}$$

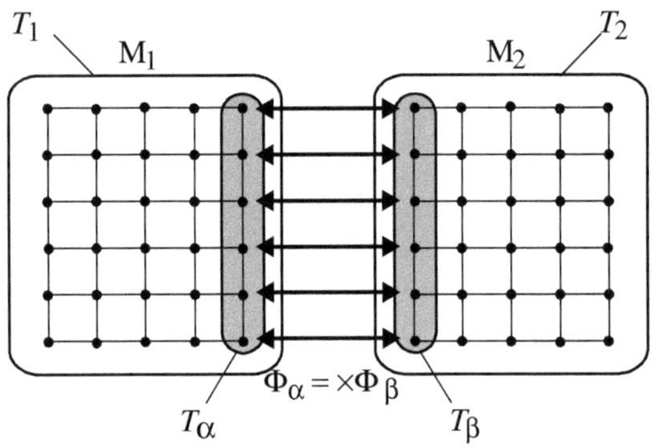

Fig. 6.43. Coupling via surface fluxes.

Let us, to begin with, suppose that M_1 and M_2 are the full-scale models with surface nodes vectors T_α and T_β. The heat fluxes between the models surfaces are introduced as

$$\begin{aligned} q_\alpha &= h\cdot(T_\alpha - T_\beta) \\ q_\beta &= h\cdot(T_\beta - T_\alpha) \end{aligned} \tag{6.50}$$

where the heat transfer coefficient h can be chosen on the basis of experimental measurements. The coupling of M_1 and M_2 can be done by adding the right-hand side of Equation 6.50 to Equation 6.49 while first rewriting Equation 6.50 into a matrix form:

$$\begin{bmatrix} q_\alpha \\ q_\beta \end{bmatrix} = H\cdot\begin{bmatrix} T_1 \\ T_2 \end{bmatrix} \tag{6.51}$$

It should be noted that H is of low rank, i.e. only the rows and columns corresponding to the coupling nodes T_α and T_β are nonzero. These rows contain h and $-h$ entries at the α_i and β_i locations:

$$H = \begin{matrix} \\ \alpha_i \\ \\ \beta_i \\ \\ \end{matrix}\begin{bmatrix} 0 & 0 & \dots & 0 & 0 \\ 0 & h & \dots & -h & 0 \\ & & & & \\ 0 & -h & \dots & h & 0 \\ 0 & 0 & \dots & 0 & 0 \end{bmatrix} \tag{6.52}$$

A coupled system is then given through

$$\begin{bmatrix} C_{11} & 0 \\ 0 & C_{22} \end{bmatrix}\cdot\begin{bmatrix} \dot{T}_1 \\ \dot{T}_2 \end{bmatrix} + \begin{bmatrix} K_{11} & 0 \\ 0 & K_{22} \end{bmatrix}\cdot\begin{bmatrix} T_1 \\ T_2 \end{bmatrix} = \begin{bmatrix} F_1\cdot u_1 \\ F_2\cdot u_2 \end{bmatrix} + H\cdot\begin{bmatrix} T_1 \\ T_2 \end{bmatrix} \tag{6.53}$$

Note that we need a strategy on how to choose an optimal h, and that in this way the temperatures in the coupling nodes T_α and T_β, unlike the Lagrange multipliers adjunction method, are not exactly equal to each other during time evolution. It is now possible to treat the coupling flux terms as additional inputs by rewriting Equation 6.53 as

$$\begin{aligned} C_{11}\cdot\dot{T}_1+(K_{11}-h_{11})\cdot T_1 &= F_1\cdot u_1+h_{12}\cdot T_2 \\ C_{22}\cdot\dot{T}_2+(K_{22}-h_{22})\cdot T_2 &= F_2\cdot u_2+h_{21}\cdot T_1 \end{aligned} \tag{6.54}$$

where $h_{11},\ldots,h_{22}$ are the matrix blocks of H. For convenience we will set Equation 6.54 in the left hand side formulation:

$$\begin{aligned} A_1\cdot\dot{T}_1 &= T_1+B_1\cdot u_1+h_1\cdot T_2 \\ A_2\cdot\dot{T}_2 &= T_2+B_2\cdot u_2+h_2\cdot T_2 \end{aligned} \tag{6.55}$$

with

$$A_1=-(K_{11}-h_{11})^{-1}\cdot C_{11}\ ,\ A_2=-(K_{22}-h_{22})^{-1}\cdot C_{22}$$

$$B_1=-(K_{11}-h_{11})^{-1}\cdot F_1\ ,\ B_2=-(K_{22}-h_{22})^{-1}\cdot F_2$$

$$h_1=-(K_{11}-h_{11})^{-1}\cdot h_{12}\ \text{and}\ h_2=-(K_{22}-h_{22})^{-1}\cdot h_{21} \tag{6.56}$$

Equation 6.55 is valid for the system of two coupled models when no MOR is performed. Let us emphasize once more that what we really want is to couple two reduced models in cases when no surface nodes T_α and T_β have been preserved. In order to be able to follow the reverse path, we project Equation 6.55 using

$$\begin{aligned} T_1 &= V_1\cdot z_1 \\ T_2 &= V_2\cdot z_2 \end{aligned} \tag{6.57}$$

and get a coupled reduced system of the form:

$$\begin{aligned} \dot{z}_1 &= V_1^T\cdot A_1\cdot V_1\cdot z_1+V_1^T\cdot B_1\cdot u_1+V_1^T\cdot h_1\cdot V_2\cdot z_2 \\ \dot{z}_2 &= V_2^T\cdot A_2\cdot V_2\cdot z_2+V_2^T\cdot B_2\cdot u_2+V_2^T\cdot h_2\cdot V_1\cdot z_1 \end{aligned} \tag{6.58}$$

where z_1 and z_2 are the generalized coordinates. It should be noted that Equation 6.58 is formally correct, regardless of the rank of the matrices h_1 and h_2 and the choice of V_1 and V_2. Hence, if we know the projection matrices and the heat transfer coefficients at the shared surfaces, Equation 6.58 couples the reduced models. The key question becomes how to find good projection subspaces, especially if the number of coupling nodes is large, as is usually the case after the spatial discretization of the thermal domain.

The authors in [178] suggest a modal approach, which is to construct the projection matrices V_1 and V_2 by using the eigenvectors of the original system matrices $C_{11}^{-1} \cdot K_{11}$ and $C_{22}^{-1} \cdot K_{22}$ of Equation 6.49. This approach offers the advantage of not having to recreate the reduced model corresponding to the new coupling conditions, but provides no guidelines on how to choose the important eigenvectors.

It is also possible to reduce the number of shared interface nodes by coarsening the mesh. In such a case, the rank of the matrices h_1 and h_2 gets low. Hence, it is convenient to apply the block Arnoldi projection to each subsystem in Equation 6.58 with the starting vectors defined as (non-zero) columns of $\left[B_1\, h_1\right]$ and $\left[B_2\, h_2\right]$. V_1 and V_2 are computed as the basis for the block Krylov subspaces:

$$K_{r1}\left[A_1, \left[B_1\, h_1\right]\right] \text{ and } K_{r2}\left[A_2, \left[B_2\, h_2\right]\right] \tag{6.59}$$

However, the reduction of shared node numbers at meshing level requires additional knowledge of the ANSYS meshing tool and makes a process nonautomatic. It is further possible to consider only a sum of columns of h_1 and h_2 by constructing the subspaces (6.59) and then to compute Equation 6.58 as written above. This is equivalent to focusing the entire heat flow between two devices through a single point.

Both of the last approaches require a sacrifice of precision at the model's surfaces, but in cases when there is only a small temperature gradient over the surface nodes, this might be an acceptable approximation. Let us conclude this section with the statement that the coupling equation, Equation 6.58 works regardless of the number of coupling nodes and that the choice of the projection subspaces requires further research.

Structure Preserving Model Order Reduction

The coupling of reduced models via substructuring, or by fluxes, was based on decoupling the array structure, reducing each single device model and then coupling the reduced models back into an array model. In this way the number of equations corresponding to the shared surface nodes was doubled. By back coupling with Lagrange multipliers during substructuring this number was reduced again, whereas by coupling via surface fluxes this was not the case. It would clearly be best not to decouple an array at all, but to still be able to reduce only its parts (different from by block Arnoldi in Sections 6.2.2). A structure-preserving model reduction technique, which has recently received a lot of attention by mathematicians [187–189], seems to offer some possibilities in this direction.

Let us assume that the interconnected system is built of two subsystems (Figure 6.44) and described through the equation system:

$$\begin{bmatrix} \dot{x}_1 \\ \dot{x}_2 \\ \dot{x}_3 \end{bmatrix} = \begin{bmatrix} A_{11} & 0 & A_{13} \\ 0 & A_{22} & A_{23} \\ A_{31} & A_{32} & A_{33} \end{bmatrix} \cdot \begin{bmatrix} x_1 \\ x_2 \\ x_3 \end{bmatrix} + \begin{bmatrix} b_1 \\ b_2 \\ 0 \end{bmatrix} \cdot u(t) \tag{6.60}$$

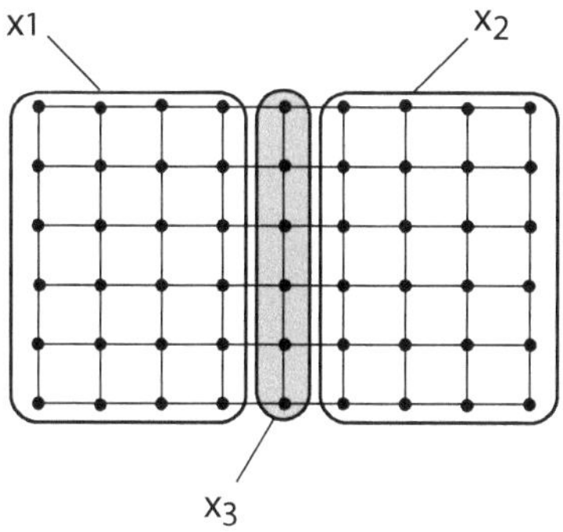

Fig. 6.44. Interconnected system whose structure should be preserved during order reduction.

For thermal systems, a diagonal heat capacity matrix was assumed in Equation 6.60. We denote the number of DOFs of both subsystems and the interface by n_1, n_2 and n_3 respectively. The goal of structure-preserving MOR is to replace Equation 6.60 with a smaller system which has the same block structure:

$$\begin{bmatrix} \dot{z}_1 \\ \dot{z}_2 \\ \dot{z}_3 \end{bmatrix} = \begin{bmatrix} A_{r11} & 0 & A_{r13} \\ 0 & A_{r22} & A_{r23} \\ A_{r31} & A_{r32} & A_{r33} \end{bmatrix} \cdot \begin{bmatrix} z_1 \\ z_2 \\ z_3 \end{bmatrix} + \begin{bmatrix} b_{r1} \\ b_{r2} \\ 0 \end{bmatrix} \cdot u(t) \tag{6.61}$$

The number of DOFs of the state vector $z(t)$ is $r_1+r_2+r_3$, with $r_1<n_1$, $r_2<n_2$ and $r_3<n_3$. Furthermore, each sub-block in Ar should be a direct reduction from the corresponding sub-blocks in the original system, e.g. A_{r11} from A_{11}. This can be accomplished by picking a block diagonal projection matrix:

$$V = \begin{bmatrix} V_1 & 0 & 0 \\ 0 & V_2 & 0 \\ 0 & 0 & V_3 \end{bmatrix} \tag{6.62}$$

and then constructing the new model with $A_r = V^T \cdot A \cdot V$ and $b_r = V^T \cdot B$.

In [188] and [189] it is stated that if V is computed by first computing a basis for the Krylov subspace:

$$span(\tilde{V}) = K_r(A^{-1}, A^{-1}b) \tag{6.63}$$

and then partitioning $\tilde{V}$ as

$$\tilde{V} = \begin{matrix} n_1 \\ n_2 \\ n_3 \end{matrix} \begin{bmatrix} \tilde{V}_1 \\ \tilde{V}_2 \\ \tilde{V}_3 \end{bmatrix} \tag{6.64}$$

and orthogonalizing the $\tilde{V}_i$'s using the QR method to get:

$$V_i = \tilde{V}_i R_i \,, \quad V_i^T V_i = I_r \tag{6.65}$$

then the transfer functions of Equation 6.60 and Equation 6.61 match in the first r moments around arbitrary frequency.

We observe a special case when the moment matching about infinity is aimed, i.e.

$$span(\tilde{V}) = K_r(A,b) \tag{6.66}$$

is computed [170]. In this case, it is not difficult to see that due to $A_{12} = A_{21} = b_3 = 0$ in Equation 6.60 we have:

$$span(\tilde{V}_1) \approx K_r(A_{11}, b_1) \text{ and} \tag{6.67}$$

$$span(\tilde{V}_2) \approx K_r(A_{22}, b_2) \tag{6.68}$$

because

$$Ab = \begin{bmatrix} A_{11} \cdot b_1 \\ A_{22} \cdot b_2 \\ A_{31} \cdot b_1 + A_{32} \cdot b_2 \end{bmatrix}, \quad A^2 b = \begin{bmatrix} A_{11}^2 \cdot b_1 + Sum \\ A_{22}^2 \cdot b_2 + Sum \\ Sum \end{bmatrix}, \text{ etc.} \tag{6.69}$$

and hence, V_1 and V_2 may be approximated directly from the submatrices A_{11} and A_{22}, i.e. from the submodels without shared interface nodes. However, it is not clear how to approximate V_3. If we assume that the number of interface nodes is to be preserved during reduction, we may construct a reduced order matrix and the load vector of the interconnected system as

$$A_r = \begin{array}{c} \\ r_1 \\ r_2 \\ n_3 \end{array} \begin{array}{c} \begin{array}{ccc} r_1 & r_2 & n_3 \end{array} \\ \begin{bmatrix} V_1^T \cdot A_{11} \cdot V_1 & 0 & V_1^T \cdot A_{13} \cdot I \\ 0 & V_2^T \cdot A_{22} \cdot V_2 & V_2^T \cdot A_{23} \cdot I \\ I \cdot A_{31} \cdot V_1 & I \cdot A_{32} \cdot V_2 & I \cdot A_{33} \cdot I \end{bmatrix} \end{array} \tag{6.70}$$

and

$$b_r = \begin{bmatrix} V_1^T \cdot b_1 \\ V_2^T \cdot b_2 \\ 0 \end{bmatrix} \tag{6.71}$$

without having to compute $K_r(A,b)$ at all, but rather computing only subspace from Equation 6.67 and Equation 6.68. Unfortunately, in this way only the moments about infinity could be approximately matched and even this is not guaranteed, for the last n_3 rows of $K_r(A,b)$ are not necessarily a unity matrix, as assumed in Equation 6.66. Hence, we can offer no proof at present for Equation 6.66. Besides, the problem of having to preserve the dimension of the interface block, which was already present in substructuring, remains. Nevertheless, this method deserves further research.

6.2.5 Overview of Proposed Strategies

We have presented several methods for a MOR of interconnected thermal systems. Presently, we are able to apply the block Arnoldi method and the Guyan-based substructuring. Block Arnoldi is straightforward, reduces the entire array model and offers high accuracy. Its main disadvantage is that it is not easily scalable to a large number of devices within an array.

Substructuring requires the decoupling of the array model and physical preservation of all shared nodes during reduction with modified Guyan's algorithm. This allows easy back coupling of the reduced models afterwards, but results in an unnecessarily large reduced array model. Hence, both methods require alternatives for use in engineering practice.

We have further discussed the possibility of finding a "mixture" of both methods, i.e. to reduce each subsystem by projection and then to couple the reduced models into an array, which could work in a general case. Equation 6.58 couples the reduced models (via surface fluxes) regardless of the number of coupling nodes and regardless of the choice of the projection subspaces. The question which remains open is how to find good projection matrices. Due to the large number of shared nodes (several hundreds), it appears that the Krylov subspace methods have reached their limits. However, by observing the numerical simulation results for the microhotplate array, we find almost no temperature gradient over the model's surface nodes. This can also be expected for other thermal MEMS models, which by design allow a heat transfer mostly in a horizontal direction. In such cases it may be acceptable to focus the heat flux on one or several points, which would allow us to consider the coupling flux terms as additional inputs. Hence, the subspaces from (6.59) could be easily computed and block Arnoldi could be applied to the reduction of each subsystem. The back coupling can be done afterwards using Equation 6.58. Figure 6.45 schematically summarizes the main characteristics of the methods discussed.

An alternative method which would not involve decoupling of the array model could be connected with structure preserving MOR. This method certainly deserves more attention.

6.3 Latest Results in Model Reduction

The last section goes beyond electro-thermal modeling although we have tried to stress applications to the heat transfer when it was possible. We give you references to the latest research papers in the rapidly developing area of model reduction.

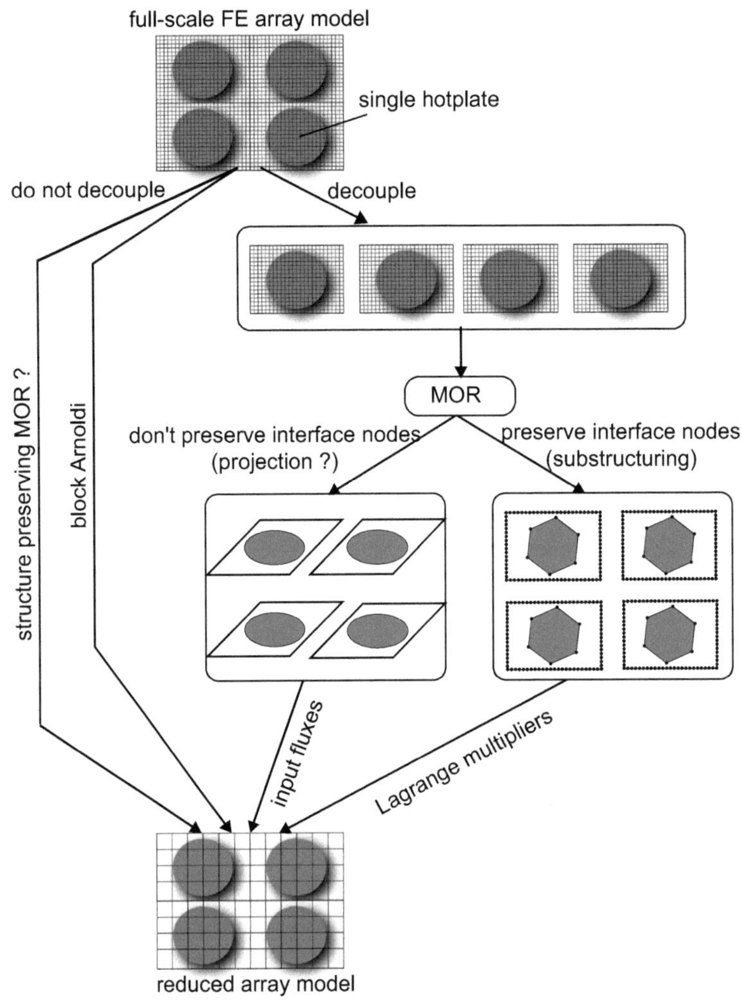

Fig. 6.45. Possibilities for MOR of interconnected systems.

6.3.1 Parametric Model Reduction

In its original form model reduction does not allow us to preserve parameters in the system matrices that naturally arise in many applications. For example, we have discussed in Chapter 2 requirements for boundary independent compact thermal models. In this case, film coefficients are within the system matrix and if they are to change one has to repeat conventional model reduction again. Fortunately, a new development, that is, parametric model reduction, allows us to overcome this limit.

To our knowledge, the first work on parametric model reduction was presented by Weile *et al.* [190] in 1999 and applied to describe frequency dependent surfaces in [191]. This approach has been generalized from two to many parameters in [192] and independently suggested in [193, 194]. We have suggested an empirical solution to a similar problem in [195] and an alternative algorithm in [196]. Note that different authors use different names for the same method: multiparameter model reduction in [192], multidimensional model reduction in [193, 194] and multivariate model reduction in [197]. Our choice in this respect was parametric model reduction as it allows us to preserve parameters in system matrices in the symbolic form.

Parametric model reduction is based on generalization of moment matching. A parameter within a system matrix looks similar to the Laplace variable in the transfer function. As a result, the idea explored by several groups was to carry out multivariate expansion of the transfer function with respect to the Laplace variable and parameters to preserve in the symbolic form simultaneously. The algorithm that generates recursively all multivariate moments up to a given order in the general case is described in [192]. However, the direct use of moments is numerically unstable and in order to employ the method in practice one has to perform orthogonalization. The method to do it in the case of one parameter was presented in [196] and comparison with previous methods was given in [198]. Codecasa *et al.* [197] presented the algorithm with the internal orthogonalization in 2004 but, in our view, the most elegant solution to this problem was found just recently in [199].

The problem to preserve film coefficients in the symbolic form was approached by several groups [196, 197, 200, 201]. The results demonstrated that the problem is solvable in principle and one can change film coefficients in the reduced model for a very wide range of numerical values while approximating well the transient behavior of the original system.

Modeling the heat transfer for a thermopile-based IR detector requires preserving the heat conductivity and heat capacity of the ambient gas in the symbolic form as the gas properties are not known in advance. In [201] parametric model reduction was used to generate a compact model containing the heat conductivity and heat capacity as parameters.

Other applications for parametric model reduction included simulation of voltammograms for a microelectrode [202], modeling a flow meter [203], interconnect [192, 193] and electromagnetism [191, 199].

The main practical problem of parametric model reduction is that the number of mixed moments grows very rapidly. For example, if we choose to preserve four parameters then a reduced model made from all first derivatives has the dimension of 6, a reduced model made from all second derivatives has the dimension 21, and a reduced model from all third derivatives already has the dimension of 56 (see Appendix F in [192]). At the same time, we may need derivatives of higher order than three to describe accurately transient behavior of the original model. A possible solution to ignore mixed moments is presented in [193, 200, 201]. Additionally in [201], the local error control has been employed to automatically choose the number of moments to include in the reduced model. However, we should state that there is no general solution to this problem yet.

6.3.2 Model Reduction for Second-Order Systems

This book was devoted to a thermal problem that after the discretization in space leads to a first-order system of ODEs. However, after the discretization of many other partial differential equations (structural mechanics, electromagnetism, acoustics, and so on), one obtains a second-order system of ODEs. Naturally, it is possible to transform it to the first-order system (see for example [167, 101]) and then perform model reduction for the transformed system.

It is interesting to note that already in 1982 Wilson [204] suggested finding a low-dimensional subspace for mechanical models by using Ritz vectors that are very similar to Arnoldi vectors. However, only Su and Craig [205, 206] have recognized the moment matching properties of Krylov subspaces. They also observed that due to a special structure of system matrix one can generate vectors in such a way as to preserve the second order in the reduced system.

Recently, the ideas of Su and Craig received attention from mathematicians. They made better proofs and hence improved the original algorithm significantly [207, 208, 209].

There is a special case of so called proportional damping when the damping matrix is a linear combination of the mass and stiffness matrices. In this case, it is possible to ignore the damping matrix during model reduction and then to form the reduced damping matrix as a linear combination of the reduced mass and stiffness matrices [210, 163].

MOR for ANSYS [12] implements three methods for a second-order system: (1) The use of a single Krylov subspace in the case of proportional damping [210], (2) Transformation to the first-order system, and (3) second-order ARnoldi (SOAR) [207].

6.3.3 Nonlinear Model Reduction

A leader among methods for nonlinear model reduction is proper orthogonal decomposition [211]. It uses simulation results of the original system ("snapshots") to build a low-dimensional subspace. It was originally developed for fluid dynamics [215] but recently has been used for structural mechanics [212, 213] and heat transfer [214].

There are recent results to generalize moment matching for a weakly nonlinear system [90, 215, 216]. With the difference to proper orthogonal decomposition, model reduction here employs system matrices directly. By a weakly nonlinear system, one understands a system in which nonlinear effects are limited to polynomial terms: quadratic, cubic, etc. The approach has recently been used for MEMS [217] and heat transfer with nonlinear film coefficient [218].

In [219, 220], the authors have used modal analysis to build a low-dimensional subspace for nonlinear MEMS models. In [221, 222, 223, 224], the nonlinear grammians have been introduced to treat nonlinearity of the original system.

An interesting idea to split a nonlinear trajectory in the pieces, build a linear reduce model for each piece, and then to merge all reduced models was explored in

[61, 225] (trajectory piecewise-linear approach). It has been used to reduce a non-linear thermal problem in [46].

6.4 Summary

Model reduction is an active area of research. More and more engineering groups are starting to use it in their own practice. At the same time, the practical importance of the problem has attracted more attention from mathematicians and they have generated new proofs and numerical algorithms. This gives us the right to state that even if your problem cannot be treated by model reduction right now, chances are high that tomorrow it will.

7

References

[1] S. D. Senturia, "Microsystem Design", *Kluwer Academic Publishers* (2001).

[2] W. Menz, J. Mohr, O. Paul, "Mikrosystemtechnik für Ingenieure" *Wiley-VCH* (2005).

[3] J. G. Korvink, E. B. Rudnyi, A. Greiner, Z. Liu, "MEMS and NEMS Simulation. In MEMS: A Practical Guide to Design, Analysis, and Applications", pp. 93–186, *Noyes Publications* (2005).

[4] A. C. Antoulas, "Approximation of Large-Scale Dynamical Systems", *Society for Industrial and Applied Mathematic* (2005).

[5] J. Lienemann, E. B. Rudnyi, J. G. Korvink, "MST MEMS model order reduction: requirements and benchmarks", *Linear Algebra and its Applications Elsevier*, 415, N 2–3, pp. 469–498 (2006).

[6] E. B. Rudnyi, J. G. Korvink, Review: automatic model reduction for transient simulation of MEMS-based devices", *Sensors Update*, **11**, pp. 3–33 (2002).

[7] C.-K. Cheng, J. Lillis, S. Lin, N. Chang, Interconnect Analysis and Synthesis", *John Wiley & Sons, Inc* (2000).

[8] A. Nathan, H. Baltes, "Microtransducer CAD Physical and Computational Aspects", *Springer Verlag* (1999).

[9] B. N. Datta, "Numerical Methods for Linear Control Systems", *Elsevier* (2004).

[10] G. Obinata, B. D. O. Anderson, "Model Reduction for Control System Design", *Springer* (2004).

[11] Z. Q. Qu, "Model Order Reduction Techniques with applications in finite element analysis", *Springer* (2005).

[12] E. B. Rudnyi, J. G. Korvink, Model order reduction for large scale engineering models developed in ANSYS. Lecture Notes in Computer Science, Vol. 3732, pp. 349–356, (2006).
http://www.imtek.uni-freiburg/simulation/mor4ansys.

[13] A. Varga, "Model reduction software in the SLICOT library", in: *Applied and Computational Control, Signals and Circuits*, Ed. B. Datta, Kluwer Academic Publishers, Vol. 2, pp. 239–282 (2001).

[14] G. Storti-Gajani, A. Brambilla, A. Premoli, "Electrothermal dynamics of circuits: Analysis and simulations", *IEEE Transactions on Circuits and Systems I-Fundamental Theory and Applications*, **48**, pp. 997–1005 (2001).

[15] M. N. Sabry, "Static and dynamic modeling of IC s", *Microelectronics Journal*, **30**, pp. 1085–1091 (1999).

[16] R. Viswanath, V. Wakharkar, A. Watwe, V. Lebonheur, "Thermal performance challenges from silicon to systems," *Intel Technology Journal* (2000).

[17] J. Adam, C. S. Chang, J. J. Stankus, M. K. Iyer, W. T. Chen, "Addressing packaging challenges - More focus on packaging research and technology development is required to handle shifting industry needs," *IEEE Circuits & Devices*, **18**, pp. 40–49 (2002).

[18] P. Tadayon, "Thermal challenges during microprocessor testing", *Intel Technology Journal* (2000).

[19] J. D. Parry, J. Rantala, C. J. M. Lasance, "Enhanced electronic system reliability - Challenges for temperature prediction", *IEEE Transactions on Components and Packaging Technologies*, **25**, pp. 533–538 (2002).

[20] H. Ye, M. H. Lin, C. Basaran, "Failure modes and FEM analysis of power electronic packaging", *Finite Elements in Analysis and Design*, **38**, pp. 601–612 (2002).

[21] N. Y. A. Shammas, "Present problems of power module packaging technology", *Microelectronics Reliability*, **43**, pp. 519–527 (2003).

[22] W. Nakayama, "Thermal issues in microsystems packaging", *IEEE Transactions on Advanced Packaging*, **23**, pp. 602–607 (2000).

[23] T. Ebefors, J. U. Mattsson, E. Kalvesten, G. Stemme, "A robust micro conveyer realized by arrayed polyimide joint actuators", *Journal of Micromechanics and Microengineering*, **10**, pp. 337–349 (2000).

[24] D. N. Pagonis, G. Kaltsas, A. G. Nassiopoulou, "Fabrication and testing of an integrated thermal flow sensor employing thermal isolation by a porous silicon membrane over an air cavity", *Journal of Micromechanics and Microengineering*, **14**, pp. 793–797 (2004).

[25] J. Wöllenstein, H. Böttner, J. A. Plaza, C. Cane, Y. Min, H. L. Tuller, "A novel single chip thin film metal oxide array", *Sensors & Actuators B: Chemical,* **93**(1-3), pp. 350–355 (2003).

[26] M. Graf, S. Taschini, P. Käser, C. Hagleitner, A. Hierlemann, H. Baltes, "Digital MOS-transistor-based microhotplate array for simulations detection of enviromentally relevant gases", *Proc. IEEE MEMS*, pp. 351–354 (2004).

[27] D. Hohlfeld, H. Zappe, "All-dielectric tunable optical filter based on the thermo-optic effect", *Journal of Optics A: Pure and Applied Optics*, **6**(6), pp. 504–511 (2003).

[28] D. Hohlfeld, "Silicon based tunable optical filters", *PhD Thesis*, University of Freiburg (2005).

[29] S. S. Yun, J. H. Lee, "Crystalline Si-based tunable Fabry-Perot filter for in-plane optical integration", *Sensors and Materials*, **15**(5), pp. 269–281 (2003).

[30] B. Courtois, J. M. Karam, M. Lubaszewski, V. Szekely, M. Rencz, K. Hofmann, and M. Glesner, "CAD tools and foundries to boost microsystems development," *Materials Science and Engineering B-Solid State Materials for Advanced Technology*, **51**, pp. 242–253 (1998).

[31] M. Strasser, R. Aigner, M. Franosch, G. Wachutka, "Miniaturized thermoelectric generators based on poly-Si and poly-SiGe surface micromachining", *Sensors and Actuators A: Physical*, **97**, pp. 535–542 (2002).

[32] J. Schilz, "Applications of thermoelectric infrared sensors (thermopiles) for remote temperature measurements" (2000). http://optoelectronics.perkinelmer.com/content/whitepapers/pyrometry.pdf.

[33] D. L. Wolffenbuttel, R. F. Dept, "Characterisation and modelling of planar on-chip integrated Peltier elements for highly localised thermal stabilisation and cooling Wijngaards", *Semiconductor Thermal Measurement and Management*, 18th Annual IEEE Symposium, pp. 105–112, (2002).

[34] http://mstmicro.com/sensor/sensor.html
[35] J. G. Korvink, H. Baltes, "Microsystem modeling", *Sensors Update*, **2**, pp. 181–209 (1996).
[36] M. Graf, R. Juriscka, D. Barrettino, A. Hierelemann, "3D nonlinear modeling of microhotplates in CMOS technology for use as metal-oxide-based sensors", *Journal of Micromechanics and Microengineering*, **15**, pp. 190–200 (2005).
[37] J. Puigcorbe, D. Vogel, B. Michel, A Vila, I. Gracia, C. Cane, J. R. Morante "High temperature degradation of Pt/Ti electrodes in micro-hotplate gas sensors", *Journal of Micromechanics and Microengineering*, **13**(4), pp. 119–124 (2003).
[38] S. M. Lee, D. C. Dyer, J. W. Gardner, "Design and optimisation of a high-temperature silicon micro-hotplate for nanoporous palladium pellistors", *Microelectronics Journal*, **34**(2), pp. 115–26 (2003).
[39] D. Briand, S. Heimgartner, M. A. Gretillat, B van der Schoot, N. F. de Rooij, "Thermal optimization of micro-hotplates that have a silicon island", *Journal of Micromechanics and Microengineering*, **12**(6), pp. 971–978 (2002).
[40] Y. Guizhen, Z. Tang, P. Chan, J. Sin, I. M. Hsing, Y. Wang, "An experimental study on high-temperature metallization for micro-hotplate-based integrated gas sensors", *Sensors & Actuators B: Chemical*, **86**(1), pp. 1–11 (2002).
[41] R. Zhang, Y. Gu, D. Zhu, J. Hao, Z. Tang, "Thermal measurement and analysis of micro hotplate array using thermography", *Sensors & Actuators A: Physical*, **100**(2-3), pp.144–152 (2002).
[42] C. Rossi, "Micropropulsion for space", *Sensors Update*, **10**, pp. 257–292 (2002).
[43] M. McNeal, N. Moelders, M. Pralle, I. Puscasu, L. Last, W. Ho, A. Greenwald, J. Daly, E. Johnson, T. George, "Development of optical MEMS CO_2 sensors", *Proc. SPIE*, **4815**, pp. 30–35 (2002).
[44] D. L. DeVoe, "Thermal Issues in MEMS and Microscale Systems", *IEEE Transactions on Components, Packaging and Manufacturing Technology*, **24**(4), pp. 576–583 (2003).
[45] L. Yin, G. K. Ananthasuresh, "A novel topology design scheme for the multi-physics problems of electro-thermally actuated compliant micromechanisms", *Sensors & Actuators A: Physical*, **97-98**, pp. 599–609 (2002).
[46] Y.-J. Yang, K.-Y. Shen, "Nonlinear heat-transfer macromodeling for MEMS thermal devices", *Journal of Micromechanics and Microengineering*, **15**, pp. 408–418 (2005).
[47] G. Lammel, S. Schweizer, S. Schiesser, P. Renaud, "Tunable optical filter of porous silicon as key component for a MEMS spectrometer", *Journal of Microelectromechanical Systems*, **11**(6), pp. 815–828 (2002).
[48] E. Peeters, S. Verdonckt-Vandebroek, "Thermal ink jet technology", *IEEE Circuits and Devices Magacine*, **13(4)**, pp. 19–23 (1997).
[49] T. K. Kun, C. J. Kim, "Valveless pumping using traversing vapor bubbles in microchannels", *Journal of Applied Physics*, **83**(11), pp. 5658–5664 (1998).
[50] P. Barth, C. Beatty, L. Field, J. Baker, G. Gordon, "A robust normally-closed silicon microvalve", *Proc. Hilton Head*, pp. 248–250 (1994).
[51] M. Freygang, H. Glosch, H. Haffner, S. Messner, "Thermal actuation units for microvalves and micropumps", *Proc. Actuator*, pp. 84–87 (1996).
[52] J. S. Fitch, A. K. Henning, E. B. Arkilic, J. M. Harris, "Pressure- based mass-flow control using thermopneumatically-actuated microvalves", *Proc. Hilton Head*, pp. 162–165 (1998).
[53] C. A. Rich, K. D. Wise, "A thermopneumatically-actuated microvalve with improved thermal efficiency and integrated state sensing", *Proc. Hilton Head*, pp. 234–237 (2000).

[54] J Palacin, M Salleras, M Puig, J Samitier, S Marco, "Evolutionary algorithms for compact thermal modelling of microsystems: application to a micro-pyrotechnic actuator", *Journal of Micromechanics and Microengineering,* **14**(7), pp. 1074–1082 (2004).

[55] http://www.ccmicro.rl.ac.uk/info_microfluidics.html

[56] M. N. Sabry, "Application of static and dynamic compact thermal models for the thermal analysis of electronic parts", *EUROSIME04 - Compact Thermal Models - short course*, (2004).

[57] Y. S. Ju, K. E. Goodson, "Microscale Heat Conduction in Integrated Circuits and Their Constituent Films", *Kluwer Academic Publishers* (1999).

[58] N. Y. Shammas, M. P. Rodriguez, A. T. Plumpton, N. Newcombe, "Finite element modelling of thermal fatique effects in IGBT modules", *IEEE Procidings on Circuit Devices and Systems*, **148**, pp. 95–100 (2001).

[59] C. Moosmann, E. B. Rudnyi, a Greiner, J. G. Korvink, "Model order reduction for linear convective thermal flow", *Proc. THERMINIC*, pp. 317–321 (2004).

[60] T. Belytschko, W. K. Liu, B. Moran, "Nonlinear Finite Elements for Continua and Structures", *J. Wiley & Sons* (2001).

[61] M. Rewienski, J. White, "A trajectory piecewise-linear approach to model order reduction and fast simulation of nonlinear circuits and micromachined devices", *IEEE Transactions on Computer-Aided Design of Integrated Circuits and Systems*, **22**, pp. 155–170 (2003).

[62] W. Batty, C. E. Christoffersen, A. J. Panks, S. David, C. M. Snowden, M. B. Steer, "Electrothermal CAD of power devices and circuits with fully physical time-dependent compact thermal modeling of complex nonlinear 3-D systems", *IEEE Transactions on Components and Packaging Technologies*, **24**(4), pp. 566–590 (2001).

[63] V. d'Alessandro, N. Rinaldi, "A critical review of thermal models for electro-thermal simulation", *Solid-State Electronics*, **46**(4), pp. 487–496 (2002).

[64] M. Pesare, a. Giorgio, A. G. Perri, "Optimized electrothermal design of integrated devices through the solution to the non-linear 3-D heat flow equation", *Microelectronics Journal*, **32**, pp. 823–831 (2001).

[65] J. V. Beck, K. D. Cole, A. Haji-Sheikh, B. Litkouhi, "Heat conduction using Green's functions", *New York: Hemisphere* (1992).

[66] A. Haji-Sheikh, J.V. Beck, D. Agonafer, "Steady-state heat conduction in multi-layer bodies", *International Journal of Heat and Mass Transfer*, **46**, pp. 2363–2379 (2003).

[67] T. Smy, D. Walkey, S. K. Dew, "Transient 3D heat flow analysis for integrated circuit devices using the transmission line matrix method on a quad tree mesh", *Solid-State Elctronics*, **45**, pp. 1137–1148 (2001).

[68] R. Hocine , A. B. Stambouli , A. Saidane, "A three-dimensional TLM simulation method for thermal effect in high power insulated gate bipolar transistors", *Microelectronic Engineering*, **65**, pp. 293–306 (2003).

[69] J. T. Hsu, L. Vu-Quoc, "A rational formulation of thermal circuit models for electrothermal simulation - Part I: Finite element method", *IEEE Transactions on Circuits and Systems*, **43**(9), pp. 721–732 (1996).

[70] C. Lasance, H. Vinke, H. Rosten, K. L. Weiner, "A novel approach for the thermal characterization of electronic parts", *Proc. SEMI-THERM IX*, pp. 1–9 (1995).

[71] V. Székely, M. Rencz, "Thermal dynamics and the time constant domain", *IEEE Transactions on Components and Packaging Technologies*, **23**(3), pp. 587–594 (2000).

[72] C. Lasance, "Recent progress in compact thermal models", *IEEE SEMI-THERM Symposium*, pp. 290–299 (2003).
[73] http://www.hitech-projects.com/euprojects/profit/
[74] H. I. Rosten, C. J. M. Lasance, J. D. Parry, "The world of thermal characterization according to DELPHI - Part I: Background to DELPHI", *IEEE Transactions on Components Packaging and Manufacturing Technology*, **20**, pp. 384–391 (1997).
[75] C. J. M. Lasance, H. I. Rosten, J. D. Parry, "The world of thermal characterization according to DELPHI - Part II: Experimental and numerical methods", *IEEE Transactions on Components Packaging and Manufacturing Technology*, **20**, pp. 392–398 (1997).
[76] H. Vinke C. J. M. Lasance, "Compact models for accurate thermal characterization of electronic parts," *IEEE Transactions on Components Packaging and Manufacturing Technology*, **20**, pp. 411–419 (1997).
[77] C. J. M. Lasance, "Two benchmarks to facilitate the study of compact thermal modeling phenomena", *IEEE Transactions on Components and Packaging Technologies*, **24**, pp. 559–565 (2001).
[78] F. Christiaens, B. Vandevelde, E. Beyne, R. Mertens, J. Berghmans, "A generic methodology for deriving compact dynamic thermal models, applied to the PSGA package", *IEEE Transactions on Components Packaging and Manufacturing Technology*, **21**, pp. 565–576 (1998).
[79] Y. C. Gerstenmaier, H. Pape, G. Wachutka, "Boundary independent exact thermal model for electronic systems", *Proc. MSM*, pp. 19–21 (2001).
[80] M. Rencz, V. Szekely, "Dynamic thermal multiport modeling of IC packages", *IEEE Transactions on Components and Packaging Technologies*, **24**, pp. 596–604 (2001).
[81] M. N. Sabry, "Dynamic compact thermal models used for electronic design: a review of recent progress", *Proc. IPACK*, pp. 1–17 (2003).
[82] Y. C. Garstenmaier, G. Wachutka, "Time depending temperature fields calculated using eigenfunctions and eigenvalues of the heat conduction equation", *Microelectronics Journal,* **32**, pp. 801–808 (2001).
[83] J. G. Korvink, E. B. Rudnyi, "Computer-aided engineering of electro-thermal MST devices: moving from device to system simulation", *Proc. EUROSIME* (2003).
[84] Y. J. Yang, C. Yu, "Extraction of heat-transfer macromodels for MEMS device", *Journal of Micromechanics and Microengineering*, **14**, pp. 587–596 (2004).
[85] L. Codecasa, D. D'Amore, P. Maffezzoni, "An Arnoldi based thermal network reduction method for electro-thermal analysis'', *IEEE Transactions on Components and Packaging Technologies*, **26**(1), pp. 186–192 (2003).
[86] L. Codecasa, D. D'Amore, P. Maffezzoni, "Compact modeling of electrical devices for electrothermal analysis", *IEEE Transactions on Circuits and Systems I-Fundamental Theory and Applications*, **50**, pp. 465–476 (2003).
[87] L. Codecasa, D. D'Amore, P. Maffezzoni, "Compact thermal networks for modeling packages", *IEEE Transactions on Components and Packaging Technologies*, **27**, pp. 96–103 (2004).
[88] T. Bechtold, E. B. Rudnyi, J. G. Korvink, "Automatic generation of compact electro-thermal models for semiconductor devices", *IEICE Transations on Electronics*, **3**, pp. 459–465 (2003).
[89] B. Shapiro, "Creating compact models of complex electronic systems: an overview and suggested use of existing model reduction and experimental system identification tools", *IEEE Transactions on Components and Packaging Technologie,* **26**, pp. 165–172 (2003).

[90] J. R. Phillips, "Projection-based approaches for model reduction of weakly nonlinear, time-varying systems", *IEEE Transactions on Computer-Aided Design of Integrated Circuits and Systems*, **22**, pp. 171–187 (2003).

[91] http://www.win.tue.nl/macsi-net/WG/WG2.html

[92] A. C. Antoulas, D. C. Sorensen, "Approximation of large scale dynamical systems; an overview", http://www-old.ece.rice.edu/~aca/Pub.html.

[93] A. C. Antoulas, "Approximation of linear dynamical systems", *Wiley Encyclopedia of Electrical and Electronics Engineering*, **11**, pp. 403–422 (1999).

[94] B. C. Moore, "Principal component analysis in linear systems: controlability, observability, and model reduction", *IEEE Transactions on Automation and Control*, **26**, pp. 17–32 (1981).

[95] B. Lohmann, "Ordnungsreduktion und dominanzanalyse nichtlinearer Systeme", *Reiehe 8: Meß-, steuerungs- und Regelungstechnik Nr. 406,* VDI Verlag (1994).

[96] J.-R. Li, "Model Reduction of Large Linear systems via Low Rank System Gramians", *PhD Thesis,* Massachusetts Institute of Technology (2000).

[97] R. W. Freund, "Reduced-order modeling techniques based on Krylov subspaces and their use in circuit simulation", in Applied and Computational Control Signals and Circuits, Vol. 1, B. N. Datta, Ed. Birkhauser, Boston, pp. 435–498 (1999).

[98] I. Troch, P. C. Müller, K.-H. Fasol, "Modellreduktion für Simulation und Reglerentwurf(Teil 2)", *Automatisierungstechnik,* R. Oldenburg Verlag, **40**(3), pp. 93–99 (1992).

[99] G. A. Baker Jr., P. Graves-Morris, "Padé Approximants", *Cambridge University Press* (1996).

[100] E. Chiproud, M. S. Nakhla, "Asymptotic Waveform Evaluation", *Kluwer Academic Publishers* (1994).

[101] Z. J. Bai, "Krylov subspace techniques for reduced order modeling of large- scale dynamical systems", *Applied Numerical Mathematics*, **43**, pp. 9–44 (2002).

[102] R. W. Freund, M. H. Gutknecht, N. M. Nachtigal, "An implementation of the look-ahead Lanczos algorithm for non-Hermitian matrices", *SIAM Journal of Scientific Computing*, **15**, pp. 313–337 (1994).

[103] R. W. Freund, "Krylov-subspace methods for reduced order modeling in circuit simulation", *Journal of Computational and Applied Mathematics*, **123**(1-2), pp. 395–421 (2000).

[104] R. W. Freund, P. Feldmann, "The SyMPVL algorithm and its applications to interconnect simulation", *Proc. IEEE Conference on Simulation of Semiconductor Proceses and Devices*, pp. 113–116 (1997).

[105] J. Cullum, T. Zhang, "Two-sided Arnoldi and nonsymmetric Lanczos algorithms", *SIAM Journal on Matrix Analysis and Applications*, **24**(2), pp. 303–319 (2002).

[106] B. Salimbahrami, B. Lohmann, T. Bechtold, J. G. Korvink, "Two-sided Arnoldi algorithm and its application in order reduction of MEMS", *Proc. 4th Mathmod*, pp. 102–108 (2003).

[107] L. M. Silveira, M. Kamon, I. Elfadel, J. White,"A coordinate-transformed Arnoldi algorithm for generating guaranteed stable reduced-order models of RLC circuits", *Computational Methods of Applied Mechanical Engineering*, **169**, pp. 377–389 (1999).

[108] A. Odabasioglu, M. Celik, L. T. Pileggi, "PRIMA: Passive reduced-order interconnect macromodeling algorithm", *IEEE Transactions on Computer-Aided Design of Integrated Circuits and Systems*, **17**, pp. 645–654 (1998).

[109] K. Gallivan, E. Grimme, P. Van Dooren, "A rational Lanczos algorithm for model reduction", *Numerical Algorithms*, **12**, pp. 33–63 (1996).

[110] A. Ruhe, D. Skoogh, "Rational Krylov algorithms for eigenvalue computation and model reduction", *Applied Parallel Computing, Lecture Notes in Computer Science*, **1541**, pp. 491–502 (1998).

[111] L. Codecasa, D. D'Amore, D. Maffezzoni, W. Batty, "Analytical multipoint moment matching reduction of distributed thermal networks", *IEEE Transactions on Components & Packaging Technologies*, **27**(1), pp. 87–95 (2004).

[112] ANSYS Release 5.6, User Manuals-Theorie Manual (1999).

[113] C. Meier, M. Emmenegger, S. Taschini, H. Baltes, J. G. Korvink, "Equivalent circuit model of resistive IC sensors derived with the box integration method", *IEEE Transactions on Computer-Aided Design of Integrated Circuits and Systems*, **18**(7) (1999).

[114] R. J. Guyan, "Reduction of stiffness and mass matrices", *AIAA Journal*, **3**(2), pp. 138 (1965).

[115] Z-Q. Qu, Y. Shi, H. Hua, "A reduced-order modeling technique for tall buildings with active tuned mass damper", *Earthquake Engineering and Structural Dynamics*, **30**, pp. 349–362 (2001).

[116] G. C. Archer, "A technique for reduction of dynamic degrees of freedom", *Earthquake Engineering and Structural Dynamics*, **30**, pp. 127–145 (2001).

[117] J. M. Bad, P. Benner, R. Mayo, E. S. Quintana-Ort, "Solving large sparse Lyapunov equations on parallel computers", *Lecture Notes in Computer Science: Euro-Par 2002 Parallel Processing*, Springer, pp. 687–690 (2002).

[118] J. M. Bad, P. Benner, R. Mayo, E. S. Quintana-Ort, "Parallel algorithms for balanced truncation model reduction of sparse systems", *Proc. PARA'04,* Springer, to appear (2004/2005).

[119] J. R. Li, J. White, "Low rank solution of Lyapunov equations", *SIAM Journal on Matrix Analysis and Applications*, **24**(1), pp. 260–280 (2002).

[120] T. Penzl, "A cyclic low-rank Smith method for large sparse Lyapunov equations", *SIAM Journal of Scientific Computing*, **21**, pp. 1401–1418 (2000).

[121] P. Benner, E. S. Quintana-Ort, G. Quintana-Ort, "State-space truncation methods for parallel model reduction of large-scale systems", *Parallel Computing*, pp. 1701–1722 (2003).

[122] J. G. Korvink, E. B. Rudnyi. Oberwolfach Benchmark Collection. In P. Benner, G. Golub, V. Mehrmann, D. Sorensen (eds), "Dimension Reduction of Large-Scale Systems, Lecture Notes in Computational Science and Engineering (LNCSE)", *Springer Verlag*, Vol. 45, pp. 311–315 (2005).

[123] Peter Benner, Volker Mehrmann, Vasile Sima, Sabine Van Huffel, and Andras Varga, SLICOT - A Subroutine Library in Systems and Control Theory, in "Applied and Computational Control, Signals, and Circuits", Vol. 1, pp. 499–539 (1999).

[124] E. Anderson, Z. Bai, C. Bischof, J. Demmel, J. Dongarra, J. Du Croz, A. Greenbaum, S. Hammarling, A. McKenney, S. Ostrouchov, and D. Sorensen. LAPACK Users' Guide, 2nd edn. SIAM, Philadelphia, PA (1995).

[125] L. S. Blackford, J. Choi, A. Cleary, E. D'Azevedo, J. Demmel, I. Dhillon, J. Dongarra, S. Hammarling, G. Henry, A. Petitet, K. Stanley, D. Walker, R. C. Whaley, ScaLAPACK Users' Guide, SIAM (1997).

[126] R. F. Boisvert, R. Pozo, K. A. Remington. The Matrix Market Exchange Formats: Initial Design. NIST Interim Report 5935 (1996).

[127] V. Rotkin, S. Toledo, The design and implementation of a new out-of-core sparse Cholesky factorization method. *ACM Transactions on Mathematical Software*, **30**, pp. 19–46 (2004).

[128] T. A. Davis, Algorithm 832: UMFPACK V4.3, an unsymmetric-pattern multifrontal method. ACM Transactions on Mathematical Software, 30(2), pp. 196–199 (2004).

[129] T. A. Davis, A column pre-ordering strategy for the unsymmetric-pattern multifrontal method, *ACM Transactions on Mathematical Software*, 30(2), pp. 165–195 (2004).

[130] R. C. Whaley, A. Petitet, J. Dongarra. Automated empirical optimization of software and the ATLAS project, *Parallel Computing*, **27**(1-2), pp. 3–35 (2001).

[131] http://nasatechnology.nasa.gov

[132] S. Jason, S. Sulinets, A. S. Curiel, M. Sweeting, "Earthquake forecast science research with a small satellite", *16th Annual USU Conference on Small Satellites* (2002).

[133] J. M. Andringa, D. E. Hastings, "A Systems Study on How to Dispose of Fleets of Small Satellites", *MS Thesis,* Massachusetts Institute of Technology (2001).

[134] J. Köhler, J. Bejhed, H. Kratz, F. Bruhn, U. Lindberg, K. Hjort, L. Stenmark, "A hybrid cold gas microthruster system for spacecraft", *Sensors and Actuators A: Physical*, **97-98,** pp. 587–598 (2002).

[135] W. A. de Groot, B. D. Reed, M. Brenizer, "Preliminary results of solid gas generator micropropulsion", *34th Joint Propulsion Conference*, AIAA-98-3225 (1998).

[136] A. D. Ketsdever, D. C. Wadsworth, S. E. Vargo, E. P. Muntz, "Design, optimization and fabrication of a free molecule micro-resistojet for microspacecraft thrust generation", http://www.stormingmedia.us, Report Number A679793 (1998).

[137] R.J. Cassady, W. A. Hoskins, "A micro pulsed plasma thruster (PPT) for the Dawgstar Spacecraft", *Proc. IEEE Aerospace Conference*, **4**, pp. 7–14 (2000).

[138] S.Marcuccio, A. Genovese, M. Andrenucci, "Experimental performance of field emission microthrusters", *Journal of Propulsion and Power*, **14**, pp. 774–781 (1998).

[139] A. P. London, A. A. Ayon, A. H. Epstein, S. M. Spearing, T. Harrison, "Microfabrication of a high pressure rocket engine", *Sensors and Actuators A: Physical*, **92**, pp. 351–357 (2001).

[140] C. Rossi, T. D. Conto, D. Estève, B. Larangot, "Design, fabrication and modeling of MEMS-based microthrusters for space applications", *Smart Materials and Structures*, **10**, pp. 1156–1162 (2001).

[141] C. Rossi, B. Larangot, T. Camps, A. Berthold, P. Q. Pham, D. Briand, N. F. de Rooij, M. Puig-Vidal, P. Miribel, E. Montané, E. Lopez, J. Samitier, "Solid propellant micro-rockets-towards a new type of power MEMS", *Proc. Nano Tech*, AIAA 2002 – 5756 (2002).

[142] D. H. Lewis, S. W. Janson, R. B. Cohen, E. K. Antonsson, "Digital micropropulsion", *Sensors and Actuators A: Physical*, **80**(2), pp. 143–154 (2000).

[143] D. Teasdale, V. Milanovic, P. Chang, K. S. J. Pister, "Microrockets for smart Dustv, *Smart Materials and Structures*, **10**, pp. 1145–1155 (2001).

[144] D. W. Younger, S. T. Lu, E. Choueiri, J. B. Neidert, R. E. Black, K. J. Graham, D. Fahey, R. Lucus, X. Zhu, "MEMS mega-pixel micro-thruster arrays for small satellite stationkeeping", *14th Annual USU Conference on Small Satellites* (2000).

[145] K. Takahashi, H. Ebisuzaki, H. Kajiwara, T. Achiwa, K. Nagayama, Design and testing of mega-bit microthruster arrays , *Proc. NanoTech*, AIAA 2002–5758 (2002).

[146] Z. You, G. Zhang, H. Li, Z. Wang, "Design, fabrication and test of MEMS propulsion with solid propellant", *4th Round Table on Micro/Nano Technologies for Space*, https://escies.org/public/mnt4/Poster14You-Zhang.pdf (2003).

[147] S. Tanaka, R. Hosokawa, S. Tokudome, K. Hori, H. Saito, M. Watanabe, M. Esashi, "MEMS-based solid propellant rocket array thruster with electrical feedthroughs", *Transactions of Japan Society of Aeronautics and Space Science*, **46**, pp. 47–51 (2003).

[148] Micropyros Final Report, http://www.laas.fr/~micropyros

[149] E. B. Rudnyi, T. Bechtold, J. G. Korvink, C. Rossi, "Solid propellant microthruster: theory of operation and modelling strategy", *Proc. NanoTech*, AIAA 2002–5755 (2002).

[150] J. Sapriel, D. Charissoux, V. Voloshinov, V. Molchanov, "Tunable acoustooptic filters and equalizers for WDM applications", *Journal of Lightwave Technology*, **20**(5), pp. 864–871 (2002).

[151] E. Dai, D. Wu, A. Xu, "Integrated acousto-optic tunable filter, equalizer, and switch in WDM communication system", *Proc. SPIE*, **3896**, pp. 403–411 (1999).

[152] A. Sneh, K. M. Johnson, "High-speed continuously tunable liquid crystal filter for WDM networks", *Journal of Lightwave Technology*, **14**(6), pp. 1067–1080 (1996).

[153] R. Le Dantec, T. Benyattou, G. Guillot, A. Spisser, J. L. Leclercq, P. Viktorovitch, D. Rondi, "Highly tunable and selective Fabry Perot filter based on InP-air Bragg mirrors for WDM applications", *Proc. SPIE*, **3632**, pp. 339–345 (1999).

[154] J. Daleiden, V. Rangelov, S. Irmer, F. Römer, M. Strassner, C. Prott, A. Tarraf, H. Hillmer, "Record tuning range of InP-based multiple air-gap MOEMS filter", *Electronic Letters*, **38**(21), pp. 1270–1271 (2002).

[155] L. H. Domash, E. Y. Ma, M. T. Lourie, W. F. Sharfin, M. Wagner, "Broadly tunable thin-film intereference coatings: active thin films for telecom applications", *Proc. SPIE*, **4989**, pp. 161–168 (2003).

[156] M. Iodicea, G. Cocorulloa, F. G. Della Cortea, I. Rendinaa, "Silicon Fabry-Perot filter for WDM systems channels monitoring"4989", *Optics Communications*, **183**(5-6), pp. 415–418 (2000).

[157] T. Seiyama, A. Kato, K. Fujushi, M. Nagatani, "A new detector for gaseous components using semiconductive thin films", *Analytical Chemistry*, **34** (1962).

[158] J. Suehle, R. E. Cavicchi, M. Gaitan, S. Semancik, "Tin oxide gas sensor using micro-hotplate by CMOS technology and in-situ processing", *IEEE-Electron Device Letters,* **14**, pp. 118–120 (1993).

[159] A. Götz A, J. Gràcia, JU. A. Plaza, C. Can, P. Roetsch, H. Böttner, K. Seibert, "A novel methodology for manufacturability of robust CMOS semiconductor gas sensor arrays", *Sensors Actuators B: Chemical*, **77**(1-2), pp. 395–400 (2001).

[160] I. Simon, N. Barsan, M. Bauer, U. Weimar, Micromachined metal oxide gas sensors: opportunities to improve sensor performance, *Sensors and Actuators B: Chemical*, **73**(1), pp. 1–26 (2001).

[161] J. Hildenbrand, "Simulation and Characterisation of a Micromashined Gas Sensor and Preparation for Model Order Reduction", *Diploma Thesis*, University of Freiburg (2003).

[162] V. Székely, M. Renz, "Thermal dynamics and the time constant domain", *IEEE Trans. on Comp. Pack. Techn.*, **23**(3), pp. 587–594 (2000).

[163] J. S. Han, E. B. Rudnyi, J. G. Korvink, "Efficient optimization of transient dynamic problems in MEMS devices using model order reduction", *J. Micromech. Microeng.*, **15**(4), pp. 822–832 (2005).

[164] T. Bechtold, D. Hohlfeld, E. B. Rudnyi, H. Zappe, J. G. Korvink, "Inverse thermal problem via model order reduction: determining material thermal properties of a microhotplate", *Proc. THERMINIC*, pp. 146–150 (2005).

[165] DOT Users Manual version 4.20 (Colorado Springs, CO: Vanderplaats Research and Development), at http://www.vrand.com/DOT.html

[166] G. H. Golub, C. F. Van Loan, "Matrix Computations", *The Johns Hopkins University Press* (1996)

[167] Z. J. Bai *et al.*, "New numerical techniques and tools in SUGAR for 3D MEMS simulation", *Proc. MSM2001*, pp. 31–34 (2001).

[168] Z. J. Bai, R. D. Slone, et al. "Error bound for reduced system model by Pade approximation via the Lanczos process", *IEEE Transactions on Computer-Aided Design of Integrated Circuits and Systems*, **18**(2), pp. 133–141 (1999).

[169] I. M. Jaimoukha, "General minimal residual Krylov subspace method for large-scale model reduction", *IEEE Transactions On Automatic Control*, **42**(10), pp.1422–1427 (1997).

[170] E. Grimme, "Krylov projection methods for model reduction", *PhD Thesis*, University of Illinois at Urbana-Champaign (1997).

[171] B. Salimbahrami, B. Lohmann, T. Bechtold, J. G. Korvink "A two-sided Arnoldi algorithm with stopping criterion and MIMO selection procedure", *Mathematical and Computer Modelling of Dynamical Systems*, **11**(1), pp. 79–93 (2005).

[172] A. Odabasioglu, M. Celik, L. T. Pileggi, "Practical considerations for passive reduction of RLC circuits", *Proc. Intenational Conference on Computer Aided Design*, pp. 214–219 (1999).

[173] M. Kammon, F. Wang, J. White, "Generating nearly optimally compact models from Krylov-subspace based reduced-order models", *IEEE Transactions on Computer-Aided Design of Integrated Circuits and Systems*, **47**, pp. 239–248 (2000).

[174] B. Salimbahrami, B. Lohmann, "A sequential reduction schema using Krylov subspace methods and truncated balanced realization", *GAMM 75th Annual Meeting*, TU-Dresden (2004).

[175] T. Penzl, "Eigenvalue decay bounds for solutions of Lyapunov equations: the symmetric case", *Systems and Control Letters*, **40**, pp. 139–144 (2000).

[176] Z. J. Bai, J. Demmel, J. Dongarra, A. Ruhe, H. van der Vorst, "Templates for the solution of algebraic eigenvalue problems: a practical guide", *SIAM*, Philadelphia (2000).

[177] A. C. Antoulas, "On the decay rate of Hankel singular values and related issues", *Systems and Control Letters*, **46**, pp. 323–342 (2002).

[178] D. Petit, R. Hachette, "Model reduction in linear heat conduction: use of interface fluxes for the numerical coupling", *International Journal of Heat and Mass Transfer*, **41**, pp. 3177–3189 (1998).

[179] X. Guo, D. Celo, D. J. Walkey, T. Smy, "The use of constant heat flow ports for thermal macro-models", *Proc. THERMICNIC*, pp. 323–328 (2004).

[180] M. von Arx, "Thermal Properties of CMOS Thin Films", *PhD thesis* 12743, ETH Zürich (1998).

[181] M. Graf, D. Barrettino, P. Käser, J. Cerda, A. Hierlemann, H. Baltes, "Smart single-chip CMOS microhotplate array for metal-oxide-based gas sensors", *Proc. IEEE Transducers*, pp.123–126 (2003).

[182] Y. Saad, "Iterative Methods for Sparse Linear Systems", *PWS Publishing Company* (1996).

[183] O. C. Zienkiewicz, “The Finite Elemente Method”, *McGraw-Hill Book Company, UK* (1977).

[184] C. Felippa, Introduction to finite element methods (ASEN 5007) - Fall 2004, Department of Aerospace Engineering Sciences, University of Colorado at Boulder, http://titan.colorado.edu/courses.d/IFEM.d/

[185] G. Strang, “Introduction To Applied Mathematics”, *Wellesley-Cambridge Press USA* (1986).

[186] N. S. Nise, “Control Systems Engineering”, John Wiley (2004)

[187] Z. J. Bai, B. S. Liao, “Towards an optimal substructuring method for model reduction”, *Proc. PARA*, (2004).

[188] R. C. Li, Z. J. Bai, “Structure-preserving model reduction”, *Proc. PARA* (2004).

[189] A, Vanderdorpe, P. Van Dooren, “Model reduction of interconnected systems”, Proc. *MTNS (2004).*

[190] D. S. Weile, E. Michielssen, E. Grimme, K. Gallivan, “A method for generating rational interpolant reduced order models of two-parameter linear systems”, *Applied Mathematics Letters*, **12**, pp. 93–102 (1999).

[191] D. S. Weile, E. Michielssen, Analysis of frequency selective surfaces using two-parameter generalized rational Krylov model-order reduction, *IEEE Transactions on Antennas and Propagation*, **49**, pp. 1539–1549 (2001).

[192] L. Daniel, O. C. Siong, L. S. Chay, K. H. Lee, J. White, A multiparameter moment-matching model-reduction approach for generating geometrically parameterized interconnect performance models, *IIEEE Transactions on Computer-Aided Design of Integrated Circuits and Systems*, **23**, pp. 678–693 (2004).

[193] P. K. Gunupudi, R. Khazaka, M. Nakhla, “Analysis of transmission line circuits using multidimensional model reduction techniques”, *IEEE Transactions on Advanced Packaging*, **25**, pp. 174–180 (2002).

[194] P. K. Gunupudi, R. Khazaka, M. S. Nakhla, T. Smy, D. Celo, Passive parameterized time-domain macromodels for high-speed transmission-line networks, *IEEE Transactions on Microwave Theory and Techniques*, **51**, pp. 2347–2354 (2003).

[195] L. H. Feng, E. B. Rudnyi, J. G. Korvink, Parametric model reduction to generate boundary condition independent compact thermal model. THERMINIC 2004, 10th International Workshop on Thermal Investigations of ICs and Systems, 29 September - 1 October 2004, Sophia Antipolis, France, pp. 281–285.

[196] L. H. Feng, E. B. Rudnyi, J. G. Korvink, “Preserving the film coefficient as a parameter in the compact thermal model for fast electro-thermal simulation”, *IEEE Transactions on Computer-Aided Design of Integrated Circuits and Systems,* **24**, pp. 1838–1847 (2005).

[197] L. Codecasa, D. D’Amore, P. Maffezzoni, A novel approach for generating boundary condition independent compact dynamic thermal networks of packages. THERMINIC 2004, 10th International Workshop on Thermal Investigations of ICs and Systems, 29 September - 1 October, Sophia Antipolis, France, pp. 305–310 (2004).

[198] L. H. Feng, Parameter independent model order reduction. *Mathematics and Computers in Simulation*, **68**, pp. 221–234 (2005).

[199] O. Farle, V. Hill, P. Ingelström, R. Dyczij-Edlinger, Multi-parameter polynomial model reduction of linear finite element equation systems, 5th MATHMOD, Proceedings. Volume 1: Abstract Volume, p. 148, Volume 2: Full Papers CD, 9 pp, February 8 - 10, Vienna University of Technology, Vienna, Austria, ISBN 3-901608-30-3 (2006).

[200] D. Celo, P. K. Gunupudi, R. Khazaka, D. J. Walkey, T. Smy, M. S. Nakhla, Fast simulation of steady-state temperature distributions in electronic components using multidimensional model reduction. *IEEE Transactions on Components and Packaging Technologies*, **28**, pp. 70–79 (2005).

[201] E. B. Rudnyi, L. H. Feng, M. Salleras, S. Marco, J. G. Korvink. Error Indicator to automatically generate dynamic compact parametric thermal models. THERMINIC 2005, 11th International Workshop on Thermal Investigations of ICs and Systems, 27 - 30 September, Belgirate, Lake Maggiore, Italy, pp. 139–145 (2005).

[202] L. H. Feng, D. Koziol, E. B. Rudnyi, J. G. Korvink. Parametric model order reduction for scanning electrochemical microscopy: fast simulation of cyclic voltammogram. Proceedings of 6th International conference on thermal and mechanical simulation and experiments in microelectronics and microsystems, EUROSIME 2005, Berlin, Germany, April 18-20, pp. 55–59 (2005).

[203] C. Moosmann, E. B. Rudnyi, A. Greiner, J. G. Korvink, M. Hornung. Parameter preserving model order reduction of a flow meter. Technical Proceedings of the 2005 Nanotechnology Conference and Trade Show, Nanotech 2005, May 8-12, 2005, Anaheim, California, USA, Vol. 3, p. 684–687 (2005).

[204] E. L. Wilson, M. W. Yuan, J. M. Dickens, "Dynamic analysis by direct superposition of Ritz vectors", *Earthquake Engineering & Structural Dynamics*, **10**, pp. 813–821 (1982).

[205] T. J. Su, R. R. Craig, "Model-reduction and control of flexible structures using krylov vectors", *Journal of Guidance Control and Dynamics*, **14**, pp. 260–267 (1991).

[206] T. J. Su, R. R. Craig, "Krylov model-reduction algorithm for undamped structural dynamics systems", *Journal of Guidance Control and Dynamics*, **14**, pp. 1311–1313 (1991).

[207] Z. J. Bai, Y. F. Su, "Dimension reduction of large-scale second-order dynamical systems via a second-order Arnoldi method", *SIAM Journal on Scientific Computing*, **26**, pp. 1692–1709 (2005).

[208] B. Salimbahrami, Structure Preserving Order Reduction of Large Scale Second Order Models, Thesis (2005).

[209] R. W. Freund, "Krylov subspaces associated with higher-order linear dynamical systems", *BIT Numerical Mathematics*, **45**, pp. 495–516 (2005).

[210] E. B. Rudnyi, J. Lienemann, A. Greiner, and J. G. Korvink, MOR for ANSYS: Generating compact models directly from ANSYS models. Technical Proceedings of the 2004 Nanotechnology Conference and Trade Show, Nanotech 2004, March 7-11, Boston, Massachusetts, USA, Vol. 2, pp. 279–282 (2004).

[211] P. Holmes, J. L. Lumley, G. Berkooz, Turbulence, Coherent Structures, Dynamical Systems, and Symmetry. Cambridge, New York: Cambridge University Press (1996).

[212] S. Lall, P. Krysl, J. E. Marsden, "Structure-preserving model reduction for mechanical systems", *Physica D-Nonlinear Phenomena*, **184**, pp. 304–318 (2003).

[213] M. Meyer, H. G. Matthies, "Efficient model reduction in non-linear dynamics using the Karhunen-Loeve expansion and dual-weighted-residual methods", *Computational Mechanics*, **31**, pp. 179–191 (2003).

[214] M. E. Kowalski, H. M. Jin, "Model-order reduction of nonlinear models of electromagnetic phased-array hyperthermia", *IEEE Transactions on Biomedical Engineering*, **50**, pp. 1243–1254 (2003).

[215] D. J. Lucia, P. S. Beran, W. A. Silva, "Reduced-order modeling: new approaches for computational physics", *Progress in Aerospace Sciences*, **40**, pp. 51–117 (2004).

[216] L. H. Feng. Review of model order reduction methods for numerical simulation of nonlinear circuits. *Applied Mathematics and Computation*, **167**, pp. 576–591 (2005).

[217] J. H. Chen, S. M. Kang, J. Zou, C. Liu, J. E. Schutt-Aine, "Reduced-order modeling of weakly nonlinear MEMS devices with Taylor-Series expansion and Arnoldi approach", *Journal of Microelectromechanical Systems*, **13**, pp. 441–451 (2004).

[218] L. H. Feng, E. B. Rudnyi, J. G. Korvink, C. Bohm, T. Hauck. Compact electro-thermal model of semiconductor device with nonlinear convection coefficient. In: "Thermal, Mechanical and Multi-Physics Simulation and Experiments in Micro-Electronics and Micro-Systems." Proceedings of EuroSimE 2005, Berlin, Germany, April 18-20, 2005, pp. 372–375 (2005).

[219] L. D. Gabbay, J. E. Mehner, S. D. Senturia, "Computer-aided generation of nonlinear reduced-order dynamic macromodels - I: Non-stress-stiffened case", Journal of Microelectromechanical Systems, **9**, pp. 262–269 (2000).

[220] J. E. Mehner, L. D. Gabbay, S. D. Senturia, "Computer-aided generation of nonlinear reduced-order dynamic macromodels - II: Stress-stiffened case", Journal of Microelectromechanical Systems, **9**, pp. 270–278 (2000).

[221] J. Hahn, T. F. Edgar, "Balancing approach to minimal realization and model reduction of stable nonlinear systems", Industrial & Engineering Chemistry Research, **41**, pp. 2204–2212 (2002).

[222] J. Hahn, T. F. Edgar, and W. Marquardt, "Controllability and observability covariance matrices for the analysis and order reduction of stable nonlinear systems", Journal of Process Control, **13**, pp. 115–127 (2003).

[223] M. Condon, R. Ivanov, "Model reduction of nonlinear systems", *Compel-the International Journal for Computation and Mathematics in Electrical and Electronic Engineering*, **23**, pp. 547–557 (2004).

[224] M. Condon, R. Ivanov, "Nonlinear systems - algebraic gramians and model reduction", *Compel-the International Journal for Computation and Mathematics in Electrical and Electronic Engineering*, **24**, pp. 202–219 (2005).

[225] D. Vasilyev, M. Rewienski, J. White, "Macromodel generation for bioMEMS components using a stabilized balanced truncation plus trajectory piecewise-linear approach", *IEEE Transactions on Computer-Aided Design of Integrated Circuits and Systems*, **25**, pp. 285–293 (2006).

8

Appendix

The software described in this Appendix is available at http://www.imtek.uni-freiburg.de/simulation/mor4ansys/. Here you will also find a link to a discussion list where you can post your questions. We also would like to attract you attention to IMTEK Mathematica Supplement (http://www.imtek.de/simulation//mathematica/IMSweb/) where you will find many useful Mathematica functions.

8.1 MOR for ANSYS

The goal of the manual is to describe the software `MOR for ANSYS` (pronounced "more for ANSYS") that generates compact models directly from ANSYS models by means of model order reduction (MOR) (see [1] for an overview). The advantages of this approach are as follows:

- ANSYS models are quite accurate. They can be created for a realistic CAD geometry of the device by taking into account small details that are important to make an accurate model.
- The software allows one to reduce the dimension of an ANSYS model significantly. In our experience [2], for thermal and mechanical problems, a reduced model of dimension less than 30 allows one to have an accurate representation of original ANSYS models up to dimension 500 000.
- Only minimal knowledge about the background theory is sufficient to run the software. The generation of a compact model is almost automatic.
- The computational cost of model reduction is relatively small. Usually, the model reduction time is comparable to the solution of a corresponding stationary problem in ANSYS. An analysis of the computational complexity is given in [3].

Let us explain what we mean by almost automatic. This means that there are a few important decisions to be taken by the engineer responsible for the simulation, namely:

- The software expects the ANSYS model to be linear. This means that if material properties depend on temperature, one has to choose an appropriate operation point to linearize the model. Note that the nonlinearity in the input function, however, **is** allowed.

- Currently, it is not possible to choose the best order of the reduced system *a priori*. A user should choose the order of the reduced system manually.

We believe that these two conditions are quite reasonable considering the huge reduction in model dimension that one can achieve by using `MOR for ANSYS`.

Note that we have research results (see publications at the MOR for ANSYS site) that will allow one to overcome these limits in future releases.

`MOR for ANSYS` is a command-line tool. It reads input files, performs model reduction and then writes output files. We will start with basic theory, then describe what inputs files are necessary and how one can prepare them in ANSYS. After than, we explain options that influence the model reduction process.

8.1.1 Basic Theory

After discretization, a system of ordinary differential equations is obtained:

$$\begin{aligned} M\ddot{\boldsymbol{x}} + E\dot{\boldsymbol{x}} + K\boldsymbol{x} &= B\boldsymbol{u} \\ \boldsymbol{y} &= C\boldsymbol{x} \end{aligned} \tag{8.1}$$

where $\boldsymbol{x}$ is the vector of unknowns that includes all degrees of freedom, M, E, K are the mass, damping and stiffness matrices, B is the input, C the output matrix, $\boldsymbol{u}$ is the vector of inputs and $\boldsymbol{y}$ is the vector of outputs. Note that our naming convention [4] is slightly different from the usual finite element convention of using C as the damping matrix. First-order systems are a special case of Eq. 8.1 when the mass matrix is zero but we use a different naming convention [4] and first-order systems are written as follows

$$\begin{aligned} E\dot{\boldsymbol{x}} &= A\boldsymbol{x} + B\boldsymbol{u} \\ \boldsymbol{y} &= C\boldsymbol{x} \end{aligned} \tag{8.2}$$

where $M = 0$ but $A = -K$.

`MOR for ANSYS` comprises two modules: the first to assemble Eq. 8.1 or 8.2 from ANSYS files (see the next section) and the second to run the Arnoldi process [5] to generate the reduced model.

The basic idea of model reduction is to find a low-dimensional subspace V that reasonably well approximates the transient behavior of $\boldsymbol{x}$, that is

$$\boldsymbol{x} = V\boldsymbol{z} + \varepsilon \tag{8.3}$$

where the approximation error ε is assumed to be small even though the number of columns of V (the dimension of $\boldsymbol{z}$) is much less the number of rows (the dimension of $\boldsymbol{x}$). `MOR for ANSYS` employs moment matching via the Arnoldi process [5] to construct V. The compact model is obtained by the projection of Eq. 8.1 as follows:

$$\begin{aligned} M_r\ddot{\boldsymbol{z}} + E_r\dot{\boldsymbol{z}} + K_r\boldsymbol{z} &= B_r\boldsymbol{u} \\ \boldsymbol{y} &= C_r\boldsymbol{z} \end{aligned} \tag{8.4}$$

where $M_r = V^T MV$, $E_r = V^T EV$, $K_r = V^T KV$, $B_r = V^T B$, $C_r = CV$. The meaning of inputs $\boldsymbol{u}$ and outputs $\boldsymbol{y}$ is the same as in Eq. 8.1 . Note that it is unnecessary to specify inputs $\boldsymbol{u}$ in order to run model reduction as the reduced system (8.4) is valid for any input. The transient and harmonic simulation of Eq. 8.4 is much faster than those of the original high-dimensional Eq. 8.1.

The approximation error clearly depends on the dimension of the vector z, generally the higher the better but it does not depend on number of outputs. The reduced model is produced by an iterative procedure. This means that if we have produced a reduced model of order 30, we can obtain all reduced models of lower order just by discarding the last columns in the projection matrix V and, as a result, the last rows and columns in the reduced matrices. This can be used in the recommended strategy to find an optimal dimension of the reduced model: generate a reduced model of a chosen maximum dimension (30 is a good starting point) and then compare the solution predicted by reduced model with the original ANSYS simulation. Note that the original ANSYS simulation it not required for the model reduction process itself.

The reduced matrices are written in the Market Matrix format [6] and can be easily read to any simulation software (Matlab, Mathematica, etc.). Each matrix is written in its own file. The files have names `base_name.matrix_name` where `base_name` can be specified with the option `-o` (by default `base_name` is equal to `mor`) and `matrix_name` follows convention of Eq. 8.1 for the second-order systems or (8.2) for the first-order systems. Additionally, file `base_name.C.names` contain text strings describing different outputs when each string is written on a separate line.

8.1.2 Preparing ANSYS files

Output Matrix

The output matrix is mostly for convenience. Model reduction during the Arnoldi process does not use the output matrix. There are two options to specify C. It is possible to have the complete output as one can restore the complete state vector by Eq. 8.3. In this case, use option `-f` and `MOR for ANSYS` will generate the output matrix C automatically by including all degrees of freedom. In this case, C will be equal to a permuted identity matrix in order to reorganize the state vector $\boldsymbol{x}$ to more natural node ordering. The information about on how degrees of freedom are written can be found in the file `mor.C.names`.

It may be more convenient to choose the output degrees of freedom explicitly. The second option allows this and you should write a file as follows. In the case, when there is only a single degree of freedom per node, you write a text file where each line contains

```
output_name node_number
```

where `output_name` is any text without spaces (it will be written to the file `mor.C.names`) and `node_number` is the number of the desired output node. You can use `node(x, y, z)` in your ANSYS script to find it.

When the number of degrees of freedom per node is more than one, the degree of freedom should be also specified.

```
output_name dof_name node_number
```

`dof_name` should be one of those listed in the header of the EMAT file (UX, UY, UZ, ROTX, ROTY, ROTZ, AX, AY, AZ, VX, VY, VZ, PRES, TEMP, VOLT, MAG, ENKE, ENDS, EMF, CURR).

Use option `-C` to specify the file name with the output degrees of freedom for `MOR for ANSYS`.

System Matrices

In the current version of `MOR for ANSYS`, the two files (FULL and EMAT) are used in order to read the system matrices. You generate them as follows:

```
/solu
allsel
antype,static
eqslv,sparse
nsubst,1
wrfull,1
ematwrite,yes
solve
fini
```

The WRFULL command has only been available since ANSYS 8. It makes ANSYS write FULL and EMAT files only and not perform a real solve. In previous versions of ANSYS the command is not available. Without this command, the time to extract the system matrices will be equal to that of the stationary solution.

If the original model is nonlinear then keep in mind that the element matrices are evaluated for the current state vector. With WRFULL this operation is quite fast as ANSYS does not perform a real solution, it only writes assembled element matrices to the binary file.

A FULL file generated for the sparse solver as shown above contains the load vector, the stiffness matrix, the Dirichlet and equation constraints. Unfortunately, no reliable way has been found to have all system matrices written in the FULL file for any case. For example, modal analysis does this for a mechanical system but not for a thermal one. As result, MOR for ANSYS uses an EMAT file to read other matrices.

The FULL file should be specifed first as follows

```
mor_for_ansys -C outdofs.txt file.full file.emat
```

or

```
mor_for_ansys -f file.full file.emat
```

when you would like to have all degrees of freedom as outputs.

Multiple Load Vectors

Input matrix B is made up of load vectors. In the simplest case, it comprises a single load vector and $\boldsymbol{u}$ is a scalar input function. In this case, you do nothing else. As model reduction does not depend on $\boldsymbol{u}$, you can take a step input function in ANSYS. This produces a load vector that can be used later on for any input function.

A compact model obtained by model reduction depends on the load vector. One can change the input function but not the load vector itself. This means that the load vector can easily be scaled but rotation is not allowed. When this is inappropriate, one can make the original model of multiple input form and run model reduction with several input vectors simultaneously. ANSYS allows one to specify several loads but unfortunately it saves only current load in the EMAT or FULL file. In order to generate several independent input vectors it is necessary to write several EMAT or FULL files. Note that an EMAT file may not contain the load vector in all cases. As result, it is recommended to use FULL files.

To use this strategy you could

```
!apply first load
!generate full and emat file as written above
/filname, second
!remove first load and apply the second load
!generate emat or full file
/filename, third
!remove the previous load and apply the third load
!generate emat or full file
...
```

Then you can run provided that the system matrices are in `file.emat`

```
MOR for ANSYS -C outdofs.txt file.full file.emat sec-
ond.full third.full
```

Several rules to keep in mind.

- The first two files must be a FULL and EMAT files as explained in the previous section. `MOR for ANSYS` read system matrices and the first load vector from them.
- Files must be consistent: they must be generated from the same model.
- If a current load vector is zero, MOR for ANSYS skips it and does not include to the matrix B. Check MOR for ANSYS output to understand the relationship between files and columns of B.

By default, load vectors are written to B as they are. The option `-c` may allow you to simplify the generation of load vectors. If it is specified, MOR for ANSYS corrects columns of B as follows. Let us assume the C numbering, that is, that the first vector is 0, the last is `NumberOfInputs - 1`. Then

```
for (int i = NumberOfInputs - 1; i > 0; --i)
```

$$\boldsymbol{b}_i = \boldsymbol{b}_i - \boldsymbol{b}_{i-1}$$

This means that one can apply the loads consecutively

```
!apply first load
!generate full and emat file as written above
/filname, second
!generate emat or full file
/filename, third
```

```
!generate emat or full file
...
```

and `MOR for ANSYS` corrects them by assuming that each generated load includes all the previous ones.

8.1.3 Performing Model Reduction

The Arnoldi process [5] takes as input two matrices, a square matrix with dimensionality equal to the state vector $\boldsymbol{x}$, and an equivalent of the input matrix with number of columns equal to the number of different inputs. Let us denote them as R and S. In the case of a single input, S is the vector. After n iterations, the Arnoldi process generates the projection matrix V with n columns and a Hessenberg matrix H such as

$$H = V^T R V \tag{8.5}$$

The Hessenberg matrix is discarded after the process is completed and the matrix V is used to make a reduced model. See Eq. 8.4 for the second-order system, for first order it is quite similar.

For the first-order system $R = A^{-1}E$ and $S = A^{-1}B$. The Arnoldi process is based on matrix vector multiplication. This means that we do not have to compute the system matrix inverse explicitly, and thus we obtain a very efficient algorithm. The matrix inverse is replaced by a solution system of linear equations. The system matrix is factorized once before the process and then the corresponding system of linear equations is solved quite efficiently by means of back substitution.

For the second-order system, there are three choices. By default, $R = K^{-1}M$, $S = K^{-1}B$ and the damping matrix is not used during model reduction and projected afterwards [9]. This method allows one to preserve the Rayleigh damping coefficients as parameters. Option - 1 transforms the second-order system to a first-order system before model reduction. Option - 2 chooses the use of the second-order Arnoldi (SOAR) algorithm [10].

There are several options to control model reduction process.

-N specifies the dimension of the reduced model. By default it is 30.

-t set up the tolerance to deflate the vector. By default it is 1e-12.

-s is to choose a solver. By default it is TAUCSlltmf.

-p is the solver parameter. By default it is `metis`.

Default values should be good enough for most models. The main problem is that it is difficult to predict what dimension of reduced model will produce the required accuracy. In our experience, a dimension of 30 is satisfactory for many thermal and structural mechanics problems but the user should always check if this is enough.

When a new vector is generated, MOR for ANSYS checks its norm. If it is less than the tolerance specified with -t option, it is deflated, that is, it is removed because it is assumed to represent a zero vector within rounding errors. When there are no more vectors to continue the Arnoldi process, it stops.

There are many solvers used to solve a system of linear equations. At present, MOR for ANSYS uses the TAUCS [7] and UMFPACK [8] libraries. Our recommendations are as follows.

For symmetric and positive definite matrices `TAUCSlltmf` with `metis` is the best choice. If the matrix is symmetric but indefinite, `TAUCSldlt` with `metis` is not bad although UMFPACK may be faster in this case. For nonsymmetric matrices use UMFPACK. If you run out of memory, choose out-of-core solvers. 4 Gb of RAM is recommended for high-dimensional models.

Choices available for a solvers are as follows.

- `TAUCSlltmf` – Multifrontal supernodal Cholesky decomposition.
- `TAUCSlltll` – Left-looking supernodal Cholesky decomposition.
- `TAUCSlltooc` – Out-of-core sparse Cholesky decomposition.
- `TAUCSllt` – Cholesky decomposition column by column (slow).
- `TAUCSldlt` – LDL^T factorization.
- `TAUCSlu` – Out-of-core sparse pivoting LU decomposition.
- `UMFPACK` – Multifrontal LU decomposition.

TAUCS solvers for symmetric matrices can take a parameter with the reordering method. Allowable values for a parameter are listed below. In our experience, `metis` is the best.

- `metis` - hybrid nested-dissection minimum degree ordering.
- `genmmd` - multiple minimum degree ordering.
- `md` - minimum degree ordering.
- `mmd` - multiple minimum degree ordering.
- `amd` - approximate minimum degree ordering.
- `treeorder` - no-fill ordering code for matrices whose graphs are trees.

More information about methods is in the library manuals [7, 8].

8.1.4 Advanced Options

Writing Original High Dimension Model

An option `-w` allows one to write the original high-dimensional system in the Matrix Market format [6] (see [4]). The specified string that follows the option `-w` is assumed to be the base name. Each matrix is written in the file `base_name.matrix_name` where `matrix_name` follows the convention of Eq. 8.1 for second-order systems or (8.2) for first-order systems. Additionally, file `base_name.C.names` contain text strings describing names of different outputs when each string is written on a separate line.

MOR for ANSYS can perform model reduction for models written in the Matrix Market format as follows

```
mor_for_ansys -MM base_name
```

Specifying Expansion Points

By default, `MOR for ANSYS` uses the expansion point zero. This can be modified with the option -x as follows

```
mor_for_ansys -C output.txt -x 1e5 file.emat
```

where the number after the option sets the expansion point. Note that if the expansion point is different from zero, the reduced model does not preserve the stationary state.

Additionally, one can perform a multi-point expansion. It is necessary to create a text file with each line containing

```
expansion_point number_of_vectors
```

The file name is then specified as an argument to the option `-x`.

Non-zero Initial State

The initial state for the transient simulation may be different from zero. In this case, one can compute the initial state for the reduced model by projecting the initial state x_0 to the low-dimensional subspace

$$z_0 = V^T \boldsymbol{x}_0 \tag{8.6}$$

This is an approximation as

$$\boldsymbol{x}_0 \neq V V^T \boldsymbol{x}_0 \tag{8.7}$$

If the difference between the initial state and its projection is too big, there are two solutions. The first is to increase the dimension of the reduced model. Second is to use the transformation

$$\boldsymbol{x}_{new} = \boldsymbol{x} - \boldsymbol{x}_0 \tag{8.8}$$

`MOR for ANSYS` supports both methods. Equation 8.6 is used during postprocessing and in this case one employs `MOR for ANSYS` as usual. In order to use Eq. 8.8, one can choose an option -0 followed by a file name that refers to a file where x_0 is written (numbers separated by white space). In this case, the reduced model will contain an additional file with an extension `init` where the components of x_0 will be written.

8.1.5 Running MOR for ANSYS: Summary

`MOR for ANSYS` is a command-line tool. A command contains options and file names.

```
mor_for_ansys [options] file_names
```

A command without arguments

```
mor_for_ansys
```

lists all the options and their defaults.

Option `-f` forces `MOR for ANSYS` to generate a complete output matrix.

Option `-o` file_name allows one to specify a base name for the reduced model. By default, it is `mor`.

Option `-0 file_name` allows us to specify the initial vector to make transformation (8.8).

Option `-c` forces `MOR for ANSYS` to correct load vectors.

Option `-w` file_name allows us to write the original ANSYS model in the Matrix Market format. By default, this file is not written. Note that for high-dimensional models it can take a lot of disk space.

Option `-C file_name` allows us to specify outputs.

Option `-t number` specifies a deflation criterion. When a new vector norm is less than this number, it is excluded. If this is the last vector, the whole process stops. By default, a deflation criterion is 10^{-12}.

Option `-s` allows one to specify a solver. By default, it is `TAUCS_lltmf`.

Option `-p` allows one to specify a solver parameter. By default, it is `metis`.

Option `-N number` sets a dimension for the reduced model. As was mentioned above, you are able to restore all the reduced models with lower dimensions. By default, it is set to 30.

Option `-x file_name` allows us to specify expansion points.

Option -1 forces the transformation of the second-order system to a first-order system.

Option -1 forces the use of the SOAR algorithm for the second-order system.

8.1.6 References

[1] E. B. Rudnyi, J. G. Korvink, *Automatic Model Reduction for Transient Simulation of MEMS-based Devices*, Sensors Update, **11**, pp. 3-33 (2002).

[2] J. G. Korvink, E. B. Rudnyi, *Model Order Reduction of MEMS for efficient computer aided design and system simulation*, Sixteenth International Symposium on Mathematical Theory of Networks and Systems, Belgium, July 5-9, 2004. http://www.imtek.uni-freiburg.de/simulation/mor4ansys/

[3] E. B. Rudnyi, J. G. Korvink. *Model Order Reduction for Large Scale Engineering Models Developed in ANSYS.* PARA'04, Workshop On State-of-the-art In Scientific Computing, Technical University of Denmark, June 20-23, 2004. http://www.imtek.uni-freiburg.de/simulation/mor4ansys/

[4] Oberwolfach Model Reduction Benchmark Collection, http://www.imtek.uni-freiburg.de/simulation/benchmark/.

[5] R. W. Freund, *Krylov-subspace methods for reduced order modeling in circuit simulation*, Journal of Computational and Applied Mathematics, 123, pp. 395–421, 2000.

[6] R. F. Boisvert, R. Pozo, K. A. Remington, The Matrix Market Exchange Formats: Initial Design, NISTIR 5935, http://math.nist.gov/MatrixMarket/

[7] S. Toledo, D. Chen, V. Rotkin, TAUCS – A library of sparse linear solvers, http://www.tau.ac.il/~stoledo/taucs/

[8] T. A. Davis, UMFPACK, http://www.cise.ufl.edu/research/sparse/umfpack/

[9] E. B. Rudnyi, J. Lienemann, A. Greiner, and J. G. Korvink. MOR for ANSYS: Generating Compact Models Directly from ANSYS Models. In Technical Proceedings of the 2004 Nanotechnology Conference and Trade Show, Nanotech 2004, March 7-11, 2004, Bosten, Massachusetts, USA.

[10] Z. J. Bai, K. Meerbergen, Y. F. Su. Arnoldi methods for structure-preserving dimension reduction of second-order dynamical systems. In: P. Benner, G. Golub, V. Mehrmann, D. Sorensen (eds), Dimension Reduction of Large-Scale Systems, Lecture Notes in Computational Science and Engineering. Springer-Verlag, Berlin/Heidelberg, Germany, 2005.

8.2 Post4MOR

The goal of the package is to make transient or harmonic simulations for a system of ordinary differential equations of first or second order in Mathematica.

8.2.1 DynamicSystem

The object represents a system of ordinary differential equations in the form:

$$\begin{aligned} \dot{x} &= Ax + Bu \\ y &= Cx + Du \end{aligned} \tag{8.9}$$

$$\begin{aligned} E\dot{x} &= Ax + Bu \\ y &= Cx + Du \end{aligned} \tag{8.10}$$

$$\begin{aligned} E\dot{x} + Kx &= Bu \\ y &= Cx + Du \end{aligned} \tag{8.11}$$

$$\begin{aligned} M\ddot{x} + E\dot{x} + Kx &= Bu \\ y &= Cx + Du \end{aligned} \tag{8.12}$$

The object contains the system matrices. Matrices M, E and D are optional, matrices A(K), B and C must be present. In addition the object contains output names for the vector y.

Constructors

`MakeDynamicSystem[matM, matE, matK,matB, matC, namesC, matD]` and `MakeDynamicSystem[matM, matE, matK, matB, matD]`	give a DynamicSystem object. `matM`, `matE`, and `matD` should be matrices or `Null`. `matK`, `matB`, and `matC` must be matrices. `namesC` is a list of strings. `matD` is optional and by default is `Null`. The second form makes `matC` equal to the identity matrix.

`MakeFirstOrderSystem[matA, matB, matC, namesC, matD]`, `MakeFirstOrderSystem[matE, matA, matB, matC, namesC, matD]`, and `MakeFirstOrderSystem[matE, matA, matB, matD]`	give a DynamicSystem object for a first order system. `matE`, and `matD` should be matrices or `Null`. `matA`, `matB`, and `matC` must be matrices. `namesC` is a list of strings. `matD` by default is `Null`. The last form makes `matC` equal to the identity matrix.
`MakeSecondOrderSystem[matM, matK, matB, matC, namesC, matD]`, `MakeSecondOrderSystem[matM, matE, matK, matB, matC, namesC, matD]` and `MakeSecondOrderSystem[matM, matE, matK, matB, matD]`	give a DynamicSystem object for a second order system. `matM`, `matE`, and `matD` should be matrices or `Null`. `matK`, `matB`, and `matC` must be matrices. `namesC` is a list of strings. `matD` is optional and by default is `Null`. The last form makes `matC` equal to the identity matrix.

`MakeFirstOrderSystem` follows the convention of Eq. 8.9 and 8.10 while `MakeDynamicSystem` and `MakeSecondOrderSystem` follow the convention of Eq. 8.11 and 8.12. The difference is in the sign of `MatrixK` (`MatrixA` = -`MatrixK`).

If dimensions of different matrices are not compatible, the constructors flag an exception.

Selectors

`MatrixM[sys]`	gives system matrix M of `sys`.
`MatrixE[sys]`	gives system matrix E of `sys`.
`MatrixA[sys]`	gives system matrix A of `sys`.
`MatrixK[sys]`	gives system matrix K of `sys`.
`MatrixB[sys]`	gives system matrix B of `sys`.
`MatrixC[sys]`	gives system matrix C of `sys`.
`MatrixD[sys]`	gives system matrix D of `sys`.
`OutputNames[sys]`	gives a list of output names of `sys`.

If a matrix is not present in the system, a selector returns a default matrix: the zero matrix for `MatrixD`; for a first-order system: the zero matrix for `MatrixM` and the unity matrix for `MatrixE`; for a second-order system: the unity matrix for `MatrixM` and the zero matrix for `MatrixE`.

Predicates

`MatrixMQ[sys]`	gives `True` if system matrix M is present in `sys` and `False` otherwise.
`MatrixEQ[sys]`	gives `True` if system matrix E is present in `sys` and `False` otherwise.
`MatrixDQ[sys]`	gives `True` if system matrix D is present in `sys` and `False` otherwise.
`FirstOrderSystemQ[sys]`	gives `True` if `sys` represent a first-order system and `False` otherwise.
`SecondOrderSystemQ[sys]`	gives `True` if `sys` represent a second-order system and `False` otherwise.
`ExplicitSystemQ[sys]`	gives `True` if `sys` represent an explicit first-order system and `False` otherwise.

Transformation Functions

`DeleteOutputs[sys, names]`	gives a new system when outputs defined in a list of names are deleted.
`AddOutputs[sys, matCadd, names]`	gives a new system with additional outputs defined by rows of `matCadd` and a list of names.
`DeleteDamping[sys]`	gives a new system without system matrix E. `sys` must be a second order system.
`AddDamping[sys, matE]`	gives a new system with matrix E inreased by `matE`. `sys` must be a second order system.
`AddDamping[sys, alpha, beta]`	gives a new system with matrix E increased by Rayleigh damping `alpha*matM+beta*matK`. `sys` must be a second order system.

`AddDampingRatio[sys, imatK]`	gives a new system with matrix K increased by `I imatK`. `sys` must be a second-order system.
`AddDamping Ratio [sys,dmprat]`	gives a new system with matrix K increased by `I 2 dmprat MatrixK`. `sys` must be a second-order system.
`ToFirstOrderSystem[sys]`	gives a new system in the form of a first-order system.
`ToExplicitSystem[sys]`	gives a new system in the form of an explicit first-order system.
`TakeSystem[sys, dim]`	gives a new system of dimension dim. This operation makes sense only when an original system has been obtained by an iterative model reduction method.

Input–Output

`ReadSystem[baseName]`	reads `DynamicSystem` from files with the base name `baseName`.
`WriteSystem[sys, baseName]`	writes a system to files with the base name `baseName`.
`ReadSystem[file, type]`	reads a system from a file written in the old format of `MOR for ANSYS 1.5`. type must be a string `FirstOrderSystem` or `SecondOrderSystem`.

The external representation of `DynamicSystem` is the same as produced by `MOR for ANSYS` and accepted in Oberwolfach Model Reduction Benchmarks.

8.2.2 SimulationResult

The object represents results of transient or harmonic simulation for `DynamicSystem`. We have an X-axis (time or frequency) and results for all outputs for an Y-axis. As such, there are four components: `XSeries`, `XName`, `YSeries`, `YNames`. `XSeries` is a vector and `YSeries` is a matrix. Names are text strings and they allow one to decide if two objects are compatible with each other.

Constructor

`MakeSimulationResult[ XSeries, XName, YSeries, YNames]`	gives a `SimulationResult` object. `XSeries` must be a vector, `YSeries` must be a matrix, `XName` must be a string and `YNames` must be a list of strings.

If dimensions of different components are not compatible, the constructor flags an exception.

Selectors

`XSeries[res]`	gives `XSeries` of `res`.
`XName[res]`	gives `XName` of `res`.
`YSeries[res]`	gives `YSeries` of `res`.
`YNames[res]`	gives `YNames` of `res`.

Predicates

`CompatibleResultQ[res1, res2, ...]`	gives `True` if all objects are compatible and `False` otherwise.

The criterion for compatibility is that all names are the same and in the same order.

Transformation Functions

`TransformResult[res, names]`	gives a new res whose `YSeries` correspond to names. res also can be a list of `SimulationResult`. Option `TransformFunction` to define `fun[y, x]`.
`Difference[res1, {listres}, ops]`	gives a list of differences between res1 and each object in listres. All `SimulationResult` must be compatible. Option is `ErrorFunction`, by default `ErrorFunction->Subtract`.

`TakeResult[res,num]`	takes first `num` values for results.

You may want to reduce the number of `YSeries` before plotting or to re-sort them. To this end, use `TransformResult`. `TransformFunction` is an option to define a function `fun[y, x]`. If it is a single function, it is applied to all `YSeries`. Alternatively, one can specify a list of pairs `{name, fun}`.

Another operation is to find a difference between different results for a difference plot. You define what the difference means by using an option `ErrorFunction`. In the case, if `XSeries` are different, `Difference` uses `Interpolation`. You can specify its options as additional arguments to `Difference`.

Plotting Function

`PlotResult[listres,ops]`	makes plots for the list of compatible `SimulationResult`. Options: `CommonTitle` for a title, `FunctionX` and `FunctionY` to modify series values, `MultipleListPlot` options.

`PlotResult` is a front end to `MultipleListPlot` to make a series of plots. Each plot contains a single output for all `SimulationResult` in `listres`.

Use `Difference` to make a difference plot.

Use options `FunctionY` and `FunctionX` to modify values before plotting. By default, they are equal to `Identity`.

Use `MultipleListPlot` options to make your plot look nicer.

Input–Output

`ReadResult[file]`	reads `SimulationResult` from file in the Matrix Market format.
`WriteResult[res,file]`	writes `res` to file in the Matrix Market format.
`WriteResult[res,file,"Table"]`	writes `res` to file as space separated values. This is convenient if you would like to use another plotting software.

`ReadResult[file,XName]`	reads `SimulationResult` from file in old format, where `XName` should be `"Time"` or `"Frequency"`.
`WriteResultOld[res,file]`	writes `res` to file in old format.
`WriteResultAsDirichlet[res,i,file]`	writes the *i*th set of res to file as ANSYS script to fix values as Dirichlet constraints.

An external represenation of SimulationResult is based on the MatrixMarket format. Additionally, file file.names contains XName and YNames.

8.2.3 Functions for Transient and Harmonic Simulation

`StationorySolution[sys, ops]`	gives a stationary solution. Options: `InputFunction`, `LinearSolve` options.
`HarmonicSolution[freq, sys, ops]`	gives `SimulationResult` for a given list of `freq`. Options: `InputFunction`, `LinearSolve` options.
`AnsysTransientSolution[time, sys, ops]`	gives `SimulationResult` for a given list time by using integrators like in ANSYS: backward Euler for a first-order system and Newmark for a second-order system. Options: `InputFunction`, `InitialState`, `TINTP`. `TINTP->{gamma, alpha, delta, theta}` allows us to change integrator behavior (see ANSYS manual).
`TransientSolution[time, sys, ops]`	gives `SimulationResult` for a given list time. Options: `Verbose`, `InputFunction`, `InitialState`, `NDSolve` options. It uses different approaches for different types of `DynamicSystem`. Use `ToFirstOrderSystem` or `ToExplicitSystem`, if `TransienSolution` fails.

`FrequencyConvergence[freq, sys, start, finish]` or `FrequencyConvergence[freq, sys, start, finish, step]`	gives an object for `LocalError` and `LocalErrorIndicator` computed for sys with dimensions from start to finish for frequency freq.
`LocalError[conv,true, ops]`	gives a list of local errors computed from the object conv obtained by `FrequencyConvergence` and a list of true values for the same frequency. Option is `ErrorFunction` (by default `Log10RelativeError`).
`LocalErrorIndicator[conv, true, ops]`	gives a list of local error estimates computed from the object conv obtained by `FrequencyConvergence`. Option is `ErrorFunction` (by default Log10RelativeError).

`HarmonicSolution` sets `XName` to `"Frequency"`. `TransientSolution` sets `XName` to `"Time"`.

By default, an input function is a unit step function for each input. You can modify this, by using an option `InputFunction`. For `HarmonicSolution`, `InputFunction` must be a list of numerical values. For `TransientSolution`, `InputFunction` can be a list of numerical values or a list of functions in two arguments: time and a state vector. `InitialState` by default is zero.

`HarmonicSolution` uses `LinearSolve` and you can use its options as additional arguments. `TransientSolution` uses `NDSolve` and you can use its options. `AnsysTransientSolution` implements integrators from ANSYS, use `TINTP` to set integrator constants.

`NDSolve` in `TransientSolution` uses only the last value in list time. It chooses integration points by itself adaptively. Use `Verbose->True` to see how many timesteps `NDSolve` has made. `AnsysTransientSolution` makes timesteps according to list time. As a result, `AnsysTransientSolution` is much faster but `TransientSolution` gives a more accurate result.

Functions `LocalError` and `LocalErrorIndicator` can be used to choose an optimal dimension for the reduced model as described in Section 6.1.

8.3 Mathlink interface to SLICOT

The package contains an interface to model reduction routines of the Control and Systems Library SLICOT:

- Balance & Truncate model reduction (BTA) AB09AD;

- Singular perturbation approximation based model reduction (SPA) AB09BD;
- Hankel norm approximation based model reduction (HNA) AB09CD.

The library is interfaced via Mathlink and you need to compile a Mathlink interface binary to use it. The precompiled binary for Windows can be found at the SLICOT site. First section describes how to compile binary, second how to use it from within *Mathematica*, third section gives the formal description of the package.

8.3.1 How to Compile the Interface

(1) Compile SLICOT or find a pre-compiled library (some are available at the SLICOT site).

(2) Compile BLAS and LAPACK (SLICOT uses them) or find pre-compiled libraries (some are available at Netlib). Performance highly depends on using optimized BLAS. Use ATLAS if you do not have the optimized BLAS for your platform (some pre-compiled libraries are at the ATLAS site).

(3) Change makefile as appropriate: specify correct location of Mathlink's library and header, specify correct location of SLICOT, LAPACK and BLAS.

(4) Run make (ignore warnings).

8.3.2 How to Use the Interface

There are two ways to use the interface binary. The simplest is when you have a binary for your local system. Then use the commad:

```
lnk = Install["fileName"]
```

Another option is to run the application on another system and connect to it from your system via the network. These two systems may have different OS.

In the latter case, the installation is a two-step procedure.

(1) Run a Mathlink application by itself at the target system.

(2) Connect to the application from your computer (you may need to choose the right protocol):

```
Install["portNumber@computerName", LinkMode_Connect,
LinkProtocol->"TCP"]
```

If everything went okay, then after either way the command

```
LinkPatterns[lnk]
```

gives a list of availalbe functions over the interface.

At the end of your work, uninstall the application:

```
Uninstall[lnk]
```

A Mathlink interface for Slicot will install three low-level functions: RunBTA, RunSPA, and RunHNA. They almost match the Slicot functions one-to-one. You access them via a high-level interface described below.

8.3.3 How to Call SLICOT

`CallSlicot[sys, dim, ops]`	gives a reduced system of dimension `dim` specified as Integer.
`CallSlicot[sys, precision, ops]`	gives a reduced system with given `precision` specified as Real.

`sys` must be `DynamicSystem` for the explicit system of first order. Use `ToExplicitSystem` if necessary.

Both functions return a list of three object: {`DynamicSystem` for the reduced system, Hankel singular values of the original system, `info`}. `info` is a list of at least two numbers. The first and second numbers are used and one optimal settings for TemporaryStorage. In the case of SPA and HNA, the third number is the dimension of the minimal realization of the original system.

If there was a failure, an exception is flagged.

Options:

`Method`	specifies the model reduction method: BTA, SPA, HNA. By default, `Method->BTA`.
`BalancingSquareRoot`	specifies whether to use the square-root (`True`) or balancing-free square-root (`False`) method. This option is in effect for BTA and SPA. By default, `BalancingSquareRoot->True`.
`Equlibration`	specifies whether to preliminarily equilibrate the original system (`True` of `False`). By default, `Equlibration->False`.
`TemporaryStorage`	specifies the amount of memory to allocate for temporary storage in the Mathlink application. The second number in `info` gives an optimal setting. By default, the minimal amount of memory is chosen.

Index

MIX
Papier aus verantwortungsvollen Quellen
Paper from responsible sources
FSC® C105338

FSC
www.fsc.org

If you have any concerns about our products,
you can contact us on
ProductSafety@springernature.com

In case Publisher is established outside the EU,
the EU authorized representative is:
Springer Nature Customer Service Center GmbH
Europaplatz 3, 69115 Heidelberg, Germany

Printed by Libri Plureos GmbH
in Hamburg, Germany